AF832085

Historical & Cultural Astronomy

Series Editors: WAYNE ORCHISTON, *University of Southern Queensland, Australia (wayne.orchiston@usq.edu.au)*
MARC ROTHENBERG, *Smithsonian Institution (retired), USA (josephhenr@aol.com)*
CLIFFORD CUNNINGHAM, *University of Southern Queensland, Australia (Cliff.Cunningham@usq.edu.au)*

Editorial Board: JAMES EVANS, *University of Puget Sound, USA*
MILLER GOSS, *National Radio Astronomy Observatory, USA*
DUANE HAMACHER, *Monash University, Australia*
JAMES LEQUEUX, *Observatoire de Paris, France*
SIMON MITTON, *St. Edmund's College Cambridge University, UK*
CLIVE RUGGLES, *University of Leicester, UK*
VIRGINIA TRIMBLE, *University of California Irvine, USA*
GUDRUN WOLFSCHMIDT, *Institute for History of Science and Technology, Germany*
TRUDY BELL, *Sky & Telescope, USA*

More information about this series at http://www.springer.com/series/15156

Steven R. Gullberg

Astronomy of the Inca Empire

Use and Significance of the Sun and the Night Sky

Springer

Steven R. Gullberg
College of Professional and Continuing Studies
University of Oklahoma
Norman, OK, USA

ISSN 2509-310X ISSN 2509-3118 (electronic)
Historical & Cultural Astronomy
ISBN 978-3-030-48365-4 ISBN 978-3-030-48366-1 (eBook)
https://doi.org/10.1007/978-3-030-48366-1

© Springer Nature Switzerland AG 2020
This work is subject to copyright. All rights are reserved by the Publisher, whether the whole or part of the material is concerned, specifically the rights of translation, reprinting, reuse of illustrations, recitation, broadcasting, reproduction on microfilms or in any other physical way, and transmission or information storage and retrieval, electronic adaptation, computer software, or by similar or dissimilar methodology now known or hereafter developed.
The use of general descriptive names, registered names, trademarks, service marks, etc. in this publication does not imply, even in the absence of a specific statement, that such names are exempt from the relevant protective laws and regulations and therefore free for general use.
The publisher, the authors, and the editors are safe to assume that the advice and information in this book are believed to be true and accurate at the date of publication. Neither the publisher nor the authors or the editors give a warranty, expressed or implied, with respect to the material contained herein or for any errors or omissions that may have been made. The publisher remains neutral with regard to jurisdictional claims in published maps and institutional affiliations.

Cover image credit: © alexeys/Getty Images/iStock

This Springer imprint is published by the registered company Springer Nature Switzerland AG
The registered company address is: Gewerbestrasse 11, 6330 Cham, Switzerland

Night sky at Machu Picchu. Watercolor by Jessica Gullberg

To Steven R. Gullberg II
A great son and research assistant who left us far too soon

Foreword

Father Bernabé Cobo, the Spanish Jesuit priest and missionary who, in the seventeenth century, wrote *Historia del Nuevo Mundo*, an authoritative and disciplined account of Inca history, culture, and religion, said of their shrines, or huacas, "there are so many of them and so many different kinds of them, it is impossible to write about all of them."

In this book on astronomy of the Inca Empire, Dr. Steven R. Gullberg doesn't contradict Cobo, but he does write about a lot of those huacas. He made a point, in fact, to visit, examine, measure, and photograph a subset of the myriad of these features, primarily the extraordinary sculptured rocks encountered in the Peruvian landscape. Some of these, like Kenko Grande and Tambomachay, are well known and routinely visited by tourists. Others are obscure and known only to specialists, like the astronomically important boulder at Q'espiwanka and the River Intihuatana far below Machu Picchu. Dr. Gullberg's pictorial inventory of these sites alone makes this book unusual and valuable.

Anyone who travels to Cusco, Machu Picchu, and other primary Inca sites in Peru will almost certainly encounter some of these monuments, will notice that they seem to be products of a common aesthetic, will be impressed by the effort and care it took to make them, and will remain puzzled by their function and meaning. The starting point for penetrating such mysteries is a search for patterns, and that search is the driver for this book.

Because these huacas are scattered and often hard to access, most visits to most of them are superficial. It is easy to sense they mean something, but because their meaning is elusive, they remain unexplained wonders that discourage more detailed study.

Knowing that historic Spanish chronicles document celestial connotations for some of the huacas and aware of modern field work to locate and analyze astronomically a few of these places, Dr. Gullberg embarked on an ambitious effort to assemble enough on-site information on enough of these places to reveal what, if any, patterns in their character might be discerned. In doing so, he has helpfully consolidated his and the work of others on these particular sites in one place. He has also correctly framed his work within Inca civilization and history.

Some aspects of antiquity will always elude us, and the detailed meaning and use of this exquisite Inca stonework and architecture may be among those unresolved issues. By collecting and describing a considerable number of the rock huacas, however, Dr. Gullberg, catalogues their primary attributes. In a search for patterns, he identifies themes that many of them seem to share—the solstice sun, the animating power of running water, the solidity and unexpected geometry of massive chiseled rock, and the social cohesion conferred by institutionalized ritual. It may be impossible to write about all of the rock huacas in Peru, but Dr. Gullberg has left few stones unturned.

Director of Griffith Observatory E. C. Krupp
Los Angeles, CA, USA

Preface

My fascination with archaeoastronomy began with the books of two of its great pioneers, Ed Krupp and Tony Aveni. This led to my graduate study of ancient astronomy at the University of Oklahoma and the research and writing of my Master's thesis regarding the Babylonian Astronomical Diaries. Subsequent to this I enrolled for doctoral study in astronomy at James Cook University, where Wayne Orchiston and Kim Malville gave me superb guidance as I developed a keen interest in the astronomy of the Incas in the Peruvian Andes.

My goal for this book is to illustrate the light and shadow effects of the Incas that I found during my field research at a representative sample of shrines, temples, and caves throughout the region of Peru that is in proximity with the Inca capitol of Cusco, and includes as well Saihuite, Tipon, the Sacred Valley, and Machu Picchu. Any discussions apply specifically to these areas but may be useful in extension when sites are examined throughout the former empire. This work is primarily intended to give illustration of the Inca's prowess with light and shadow to those not having visited these locations and to those who would like to do so.

It is imperative that any archaeoastronomical data collected be placed into cultural context; therefore I begin with chapters regarding Inca history and culture. This is followed by an overview of archaeoastronomy and a survey of much of what is understood of Inca astronomy. This all precedes the presentation and discussion of my research data in order to help any reader follow the processes and cultural factors involved.

My research presented in Chapters 7, 8, and 9 covers an array of sites in the designated areas. Some of my colleagues have published papers in more detail about specific sites, but this book is intended to highlight overall Inca prowess with light and shadow effects through photos and description of numerous locations. Those desiring more detail can seek out these other papers, and it is my hope that some will want to perform additional research themselves.

Photo documentation of archaeoastronomical events is imperative, and I have included many images from among the thousands that I took during my research to aid in illustration. In addition to my research photos it was my privilege to be able to collaborate with my wife, Jessica, a wonderful artist. She contributed many paintings and drawings to help me enhance the book and it was a joy to be able to work together in our effort to provide readers with greater insight. My goal was to significantly share my experiences with Inca effects of light and shadow and use these images to take readers to the world that I encountered. I hope that I have "painted" a great image that you can use to better appreciate these aspects of the Inca Empire.

I include descriptions and photos of what was found. Data such as azimuths are included in the appendices. My intention is for this is to be of service to others as they continue with their own research and draw conclusions of their own.

The final chapter summarizes my findings and the degree of support for my initial hypotheses. Results are presented in tables and graphs to make it easier for readers to see overall what was found. Here also is where I discuss some thoughts and concluding remarks.

The time that I spent in the Peruvian Andes has been one of the greatest experiences of my life. Not only was I able to examine extant Inca astronomy firsthand, but I also had the great pleasure of meeting and interacting with many of the wonderful people of Peru, both in the rural areas of my research and in Cusco as well. I hope you enjoy what is to follow.

Norman, OK, USA Steven R. Gullberg

Acknowledgments

First, I would like to both introduce and thank Milton Rojas Gamarra. Milton is Quechua and a professor at the Universidad Nacional de San Antonio Abad del Cusco. He is an outstanding scholar of Inca astronomy. Milton's insight has been instrumental, and I very much look forward to collaborating with him in the future. This has long been his passion too, and he will make most valuable contributions to the world's understanding of the astronomy of his ancestors, the Incas.

I owe a great deal to Wayne Orchiston for his guidance from the beginning of my research and throughout. I am very grateful to Kim Malville for all he taught me, especially when he accompanied and tutored me on my first research trip to the Andes. Tom Zuidema reviewed my early work and from his vast experience made significant suggestions that I value greatly. I also must thank Bernard Bell and Kenneth Wright for the inputs they gave to enhance my research.

I cannot express enough gratitude to my wife, Jessica, a gifted artist. She assisted with field research and her collaboration with the numerous watercolor paintings, pencil drawings, and maps she created greatly enhances the book throughout.

My primary research assistant on each trip was Carlos Aranabar, and I could not have completed this effort without his assistance. I thank my sons, Steven and Gregory, who while college students each assisted me on separate expeditions. Jesus Villafuerte was an important contributor to my research at Llactapata and I owe Mike Zawaski great thanks for the theodolite data that he helped me collect.

Contents

Chapter 1
Introduction

Fig. 1.1 Temple of the Condor at Machu Picchu. (Pencil by Jessica Gullberg)

This book explores Inca astronomy and its cultural context. Numerous Inca shrines have been found to exhibit symbolic representations of astronomical knowledge and meaning. Through the study of these rock carvings and their orientation, as well as

© Springer Nature Switzerland AG 2020

S. R. Gullberg, *Astronomy of the Inca Empire*, Historical & Cultural Astronomy,
https://doi.org/10.1007/978-3-030-48366-1_1

Fig. 1.2 Theodolite at Q'espiwanka. (Watercolor by Jessica Gullberg)

the observation of the Sun's light and shadow effects on significant days of the Inca calendar, this astronomical function can readily be seen.

Many relevant interpretations – astronomical, abstract sculpture, ritual, divination, worship, and ceremonial – can be gleaned from such shrines, showing the multiple levels of meaning that the Incas attached to significant landscape features. Examination of them must be holistic, considering cultural motifs, their topographic and astronomical contexts, sightlines to other shrines or prominent features of the landscape, and light and shadow effects throughout the year, especially on ceremonial or agriculturally important dates. Astronomical interpretations cannot be separated from the overall cultural significance of these objects (Fig. 1.2).

This book explores four primary areas of interest within Inca astronomy and culture. We shall set the stage with a very brief introduction to Inca worship and a summary of each point.

The Four Points of Investigation

For millennia, the people of Andean cultures preceding the Inca Empire believed the world to be filled with animate beings, some of which were ancestors with powers to influence the living. The great snow mountains, rivers, springs, rocky outcrops, or other more subtle features of the landscape could be alive and powerful and often required offerings. People interacted with these animate beings, who could be supplicated, consulted with the help of an oracle, battled, abducted, and even destroyed.

These sites formed a system of shrines, which in Quechua were known as *huacas*. At the time of the Inca Empire and Spanish conquest, there were many hundreds of them. The most powerful huacas required care and maintenance, which included gifts made to the powers of the shrines. Sacrifices were offered to them and those of animals or produce were often used to support the huaca's attendants. The Incas organized the administration of huacas along lines called *ceques*, and the existence of such a system has been well documented for the area that surrounds Cusco.

It is thought that the design and carving of Inca rock huacas followed an established tradition that included cosmology, astronomy, and mythology. Meaning was encoded into ceremonial structures in the form of intentional light and shadow effects, features framing or directing the eye to distant points on the horizon, and elements relating to origin myths such as caves, basins, and fountains. The carving and shaping of huacas were not spontaneous and were instead carefully planned in accordance with a specific iconic repertory with shared meaning and purpose.

Stone huacas were frequently elaborate, as is shown with examples in Chaps. 7, 8, and 9. The Incas revered and venerated large rocks, and their emperor, Pachacuti, apparently believed he could "improve" upon the work of his co-creator and father – the Sun – by their sculpting. As a direct result, these carved outcrops retain a great amount of information about the sacred and ritual world of the Incas. Susan Niles (1987) suggests that a fairly limited number of motifs were used in the shaping of huacas: "The pattern does not suggest a tolerance of innovation. It is unlikely that the Incas would encourage individual graffiti artists to practice their skills on sacred rocks. Certainly, the improvements on nature seen in Inca shrines...must have been officially controlled" (pp. 204–205). An element of Pachacuti's architectural style was to incorporate carefully fitted stone masonry into natural outcrops, such as at Machu Picchu, where "building seemed to grow organically out of the bedrock.... and the boundary between the work of the architect and the Creator is blurred" (Niles 2004, p. 62).

Since the Sun was the primary deity of the Incas, it is logical that many of these huacas would be associated with solar worship. This brings us to the first point.:

> *The majority of the currently identified rock huacas in the greater Cusco Valley region, the Sacred Valley, and surrounding Machu Picchu are associated with visual solar phenomena.*

Part of Inca claims for legitimacy as rulers were based upon the assertion that the Inca royalty were direct descendants of the Sun. Demonstrations supporting that genealogical link could occur during public festivals when the Sun would rise on the horizon at a location predicted by the Inca king and priests, the most likely times being identifiable elements of the solar cycle. Accordingly, the next main area explored is:

> *Those huacas found to be associated with visual solar phenomena (solar huacas) exhibit alignments related to the solstices, equinoxes, zenith Suns, and anti-zenith Suns.*

A long-standing Andean tradition involved the concept of *camay,* whereby inanimate objects were brought to life and made sacred and powerful by such as flowing water. The huacas that were energized with these waters were thought to be prominent living entities. While all huacas were not associated with water, solar huacas were largely adjacent to water courses such as stone lined or carved channels. The third main point is:

Solar huacas are associated with flowing water.

Part of the perceived power of huacas may have been derived from the inclusion of long standing Andean sacred symbols such as stairways for shamanic ascent and descent, caves, niches for sacred objects, seats for spirits, sacred animal carvings, channels for ceremonial fluids, fountains, and basins for viewing reflections in liquids of the Sun, Moon, and stars. The majority of huacas consisted of natural features, but those made of rock could be enhanced through carving. The fourth main point is:

Solar huacas in the form of carved rocks contain traditional Andean motifs.

The chapters ahead explore many aspects of the astronomy of the Inca Empire. Chapters 2, 3, and 4 provide the insight and context necessary to properly understand cultural influences upon Inca astronomy. Chapter 5 describes elements of archaeoastronomy, such as the motions of objects in the sky, and in Chap. 6 you will explore Inca astronomy in detail. These chapters will help you to more thoroughly perceive the astronomical light and shadow effects that are presented in Chaps. 7, 8, and 9. Chapter 10 then summarizes this all with discussion, charts, and graphs, and research data are presented in the appendices.

As you read, you will learn about a South American empire with an impressive astronomy. My goal is to paint a clear picture of the prowess with which the Incas were able to create light and shadow effects, and as well the great extent to which astronomy was woven throughout their empire. Next, in Chap. 2: Evolution of an Empire, you will learn about the history of this great culture.

Chapter 2
Evolution of An Empire

Fig. 2.1 Llama at Machu Picchu. (Watercolor by Jessica Gullberg)

© Springer Nature Switzerland AG 2020

S. R. Gullberg, *Astronomy of the Inca Empire*, Historical & Cultural Astronomy,
https://doi.org/10.1007/978-3-030-48366-1_2

By the time of the sixteenth century invasion from Europe the Inca Empire had grown to span the Andes and coastal regions from Argentina and Chile through Bolivia and Peru to Ecuador and into southern Columbia (Fig. 2.2). The reign of the imperial Incas was relatively short, but they were preceded by many civilizations that traced back thousands of years, and the roots of Andean astronomy can largely be attributed to those ancient cultures. Celestial myths, beliefs, and knowledge developed through observations made over the centuries, passed on from one generation to the next and from one tribe to another. The Incas inherited this celestial knowledge and adapted it to their own needs. They created a strong state religion of the Sun founded upon astronomy assimilated through their conquests. Temples and shrines were built that publicly displayed many of these astronomical concepts.

Systematic celestial development essentially had come to a halt by 1532 when fortune-seeking Spanish conquistadors invaded Peru, and not much had advanced during the internal strife of the 5 years immediately preceding the conquest. The Spaniards, however, failed to appreciate local astronomical knowledge and instead persisted in framing their views according to those of European culture.

It took the Spanish 40 years to track down and execute the last ruling Inca. Catholic priests from Spain descended upon Peru and began to locate and destroy everything related to the indigenous religion. This included all they could identify that related to worship of the Sun, Moon, and stars.

The long history of the Andes and its many social pressures greatly affected the astronomy of the Incas. Some of those influences are outlined in the sections of this chapter.

2.1 Before the Incas

The civilization of the Incas was built upon cultures that preceded them, such as the Huari, Nasca, and Chavin. Their period of dominance, lasting only a century, was far too short to independently develop such advanced forms of religion, social structure, construction, and astronomy. The Incas were adept assimilators of both the knowledge available at the time of their empire's inception, and as well as that of the many tribes conquered during their reign (D'Altroy 2002; Gasparini and Margolies 1980; Paternosto 1989).

Silverman (2004), Isbell and Vranic (2004), Haas and Creamer (2004), Solis(2006), Campbell and Rick (2004), and Ghezzi (2006) have discussed periods of Andean chronology that update earlier delineations devised by Rowe (1946). Table 2.1 is a compilation of such cultural periods. It must be stressed that exact dates remain uncertain and the ones presented simply serve as relative examples. Representative cultures are provided for context, but this list is not complete.

Fig. 2.2 The Inca Empire and its road system. (Manco Capac / CC BY-SA / https://creativecommons.org/licenses/by-sa/3.0)

Table 2.1 Andean chronology. (Author's table)

Period	Timeframe	Representative cultures
Late Horizon	1476 CE – 1532 CE	Inca
Late Intermediate	1000 CE – 1476 CE	Chimú, Chincha, Inca
Middle Horizon	550 CE – 1000 CE	Huari, Tiwanaku
Middle Intermediate	300 CE – 600 CE	Huari, Nasca
Early Intermediate	100 CE – 300 CE	Moche, Nasca
Epiformative	200 CE – 100 CE	
Final Formative	400 BCE – 200 BCE	
Late Formative	600 BCE – 400 BCE	Chavin
Middle Formative	1000 BCE – 600 BCE	Chavin
Early Formative	1500 BCE – 1000 BCE	
Final Archaic	2000 BCE – 1500 BCE	
Late Archaic	3000 BCE – 1800 BCE	Caral

Fig. 2.3 Moche pottery. (Jean-Pierre Dalbéra from Paris, France / CC BY / https://creativecommons.org/licenses/by/2.0)

More than two millennia ago in the Middle Formative, much of Peru was influenced by the Chavin; what we know of them largely comes from the art found in their archaeological remains (Burger 1992). In the Early Intermediate, to the north lived the Moche, known for their pottery and textiles (Fig. 2.3). The Huari followed in the Middle Intermediate and Middle Horizon, and later the Chimú in the Late Intermediate. In the region of southern Peru in both the Early and Middle Intermediate were the Nasca, who produced pottery and textiles as well as enigmatic lines across large landscapes, which were thought to be astronomically related and whose designs were fully discernable only from great heights (Fig. 2.4). This southern area later saw the Chincha in the Late Intermediate (Hemming 1970; Paternosto 1989; Silverman 2004 ; Zuidema 1964).

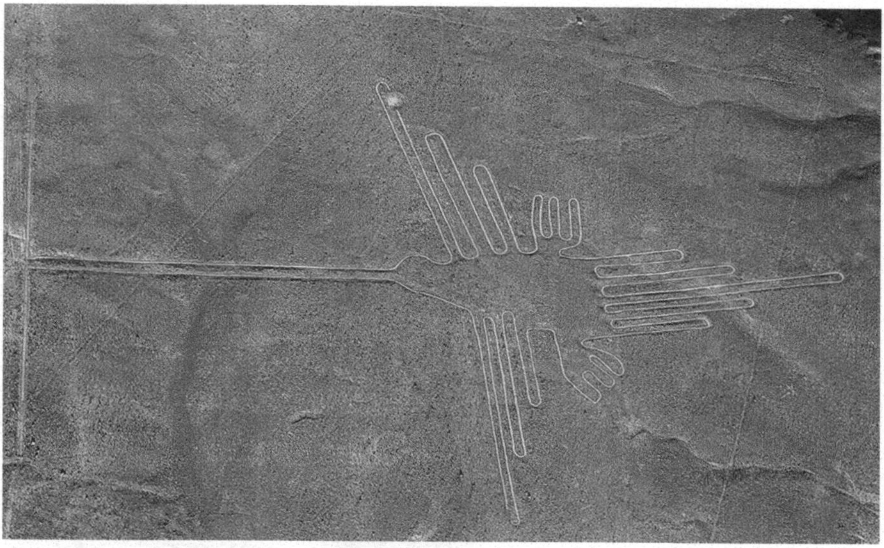

Fig. 2.4 The Hummingbird, Nasca Lines. (Diego Delso / CC BY-SA 4.0 / https://creativecommons.org/licenses/by-sa/4.0)

In the Middle Horizon, much of Peru was dominated by the Huari and the Tiwanaku. The Huari developed in Peru's Ayacucho region while the Tiwanaku stemmed from Lake Titicaca in Bolivia. The Tiwanaku occupied southern Peru, and its ruins include stone platforms, statues, and a gateway of the Sun (Fig. 2.5). The Incas, also with cosmological beginnings at Lake Titicaca, likely learned much from the Tiwanaku with regard to their construction practices (Hemming 1970).

During the Late Intermediate other cultures arose, such as the Canary, Chanca, Colla, Lupaca, Huanca, Conchuco, Yarivilca, Chachapoya, and the Incas. These tribes flourished until they each were conquered during the consolidation period of the Incas. Prior to their conquest the Chimú developed a society especially sophisticated in art, construction, irrigation, defense, and politics (Fig. 2.6). The Incas appear to have closely studied advancements in Chimú civilization (Hemming 1970).

The Incas owed a great deal to their predecessors as much of their culture and technology came from knowledge assembled over many centuries. Technique and motivation for monumental sculpture seem likely to have taken root with the Tiwanaku and Huari. The Huari also leave an intricate textile calendar. Construction with large blocks was taken from the Tiwanaku region at Lake Titicaca. The Tiwanaku shared a similar cosmology that provided the Incas with the ancestral roots they needed to establish the legitimacy of their empire (Zuidema 1977; Hemming and Ranney 1982).

An impressive site is at Chankillo (Fig. 2.7). Ruggles and Ghezzi (2006) examined Chankillo and found alignments for both the Sun and the Moon. They assert the site to be both an observatory and a calendar created some 2300 years ago.

Fig. 2.5 Tiwanaku carved stone head. (Dennis Jarvis from Halifax, Canada / CC BY-SA / https://creativecommons.org/licenses/by-sa/2.0)

Fig. 2.6 Chimú tapestry shirt. (The Michael C. Rockefeller Memorial Collection, Bequest of Nelson A. Rockefeller, 1979 / Public domain)

Fig. 2.7 June solstice sunrise over Chankillo. (Watercolor by Jessica Gullberg)

2.2 The Early Incas

> At the beginning of time the one who gave the world the vital breath was called Wiracocha, and there, where all was darkness, he placed a race of giants whom he instructed to live in peace so that they might serve him and know him for all time. (Salazar and Salazar 2014, p. 20)

It is thought that the earliest Incas settled in the Cusco valley circa 1200 CE. Their existence appears to have been relatively stable until the early fifteenth Century, a time at which many significant events in Inca history began to occur. Inca mythology proclaims Manco Capac (Fig. 2.8) to be their founding leader, descended from the god Wiracocha (Salazar and Salazar 2014). (Wiracocha is also spelled Viracocha.) Manco Capac began an unbroken dynastic succession that led to the Incas' 8th emperor, Viracocha Inca (Fig. 2.9), and his son, Inca Yupanqui, or Pachacuti.

The six emperors between Manco Inca and Viracocha Inca were: Sinchi Roca, Lloque Yupanqui, Mayta Capac, Capac Yupanqui, Inca Roca, and Huahar Huacac. Niles suggested that Manco Capac, Sinchi Roca, Lloque Yupanqui, and Mayta Capac might have been mythical since no evidence has been found to support that their mummies ever existed (D'Altroy 2002; Niles 1987). This is still up for debate. The portraits of the early Incas were painted after the Spanish conquest.

Fig. 2.8 Manco Capac. (Public domain / https://commons.wikimedia.org/wiki/File:Manco_C%C3%A1pac_18th_century_Cusco_School.jpg)

The Incas promulgated the belief that they were the chosen people of the Sun and that they had been "created to be the rulers" (Sherbondy 1992, p. 54). "They were the children of the sun and therefore first among all peoples" (Sherbondy 1992, pp. 55–56). The Incas developed their culture and religion around the Sun and thrived. Legend has it that a major change in fortune occurred c. 1438 when a powerful neighbor, the Chancas, attacked the Incas with the aim of conquest. Viracocha Inca, sensing defeat, fled Cusco with his designated heir, leaving its defense to a younger son, Inca Yupanqui. In an epic battle he routed the Chancas and following his victory he usurped rule from his father and brother (D'Altroy 2002 ; Zuidema 1964). He adopted the name Pachacuti Yupanqui Inca (Fig. 2.10) and empowered by his Chancan victory embarked on a campaign first to conquer the Chanca's allies, and then to expand Inca rule over all known tribes throughout the Andes. According to legend, Pachacuti's father never returned to Cusco (Niles 1987). Zuidema (1985) points out that no archaeological evidence has been found to support the occurrence of this battle.

Fig. 2.9 Viracocha Inca. (Brooklyn Museum / Public domain / https://www.brooklynmuseum.org/opencollection/objects/2304)

Fig. 2.10 Pachacuti Yupanqui Inca statue in Aguas Calientas. (Author's photograph)

The ensuing years were filled with conquest, but also construction and enlightenment. Pachacuti commanded a great building program that created a majority of the huacas, palaces, temples, and Inca roads that we find today. Tribes, such as the Chimú, when defeated were often dispersed throughout the empire as workers called *mitmaes* in an effort to diffuse threats to Inca security and gather specialists for service to the state (D'Altroy 2002). The Incas assimilated much from many tribes as they conquered them, and further refined their civilization.

It is often mentioned that the Tiwanaku inspired the Incas' tradition of monumental sculpture (D'Altroy 2002 ; Paternosto 1989). There is a chronological gap between these two cultures, but distinct similarities exist between their architectural styles. Additionally, as previously mentioned, in-depth knowledge of astronomical cycles can only be observed over long periods of time, thus implying that the Incas acquired much of their celestial prowess through information gathered by other civilizations, either direct predecessors or conquered tribes holding ancient traditions.

Pachacuti, his son, and his grandson successively built the largest empire ever known in the pre-Columbian Americas, 4800 km from Chile to Columbia. They established armies in fortresses at distant stations with the infrastructure of roads and storehouses necessary to support them. Temples and shrines were constructed as a part of exerting state control over subjects, as well as pilgrimage centers designed to reinforce the legitimacy of royal rule over the populace (Bauer and Stanish 2001).

By the time Francisco Pizarro arrived in Peru in 1532, the Incas had not only built an impressive empire, but also a remarkable society. Their civilization had advanced to its sophisticated level without European influence or even knowledge that nations such as Spain existed (Hemming 1970). As far as "the son of the Sun" was concerned, he ruled the world.

2.3 Rulers of Imperial Expansion

The greatest period in Inca history began with Pachacuti's ascent and lasted until the c. 1527 death of his grandson, Huayna Capac. Pachacuti's son, Topa Inca, governed from 1471 to 1493 (Niles 1987). Pachacuti's ambition and vision launched an empire of a magnitude never before seen in the Western Hemisphere. He initiated massive construction, created a complex society and bureaucracy, further developed a domineering religion, and had very apparent interests in celestial alignments. His son and grandson continued to pursue his visions of imperial expansion and extensive construction, albeit building with interpretations of their own. It would take civil war and a European invasion to halt their progress.

Pachacuti Yupanqui Inca

Pachacuti, the name adopted by Inca Yupanqui, translates as "transformer of the world," which was exactly his intent. More specifically, from Quechua, *pacha* means "a moment or interval in time and a locus or extension in space" (Salomon and Urioste 1991, p. 14) and *cuti* "to turn around" (Hemming and Ranney 1982; Fig. 2.10). The ninth king of the Incas viewed himself both as the son of the Sun and co-creator of the land. Salazar (2004) suggests that Pachacuti felt a special association with "supernatural forces imminent in the landscape and the celestial sphere," and that his connection with these forces needed to be "actively reaffirmed through daily ritual" (p. 41).

The Incas first sought to control the Chancas. They later set out to subjugate the world, including the region surrounding Lake Titicaca (Fig. 2.11) – that of their mythical origins. Pachacuti aggressively expanded the Incas' sphere by conquering one tribe after another. He desired to spread Inca religion and culture across the Andes (Niles 1987). Noteworthy is that while the Incas were known for their military prowess, Pachacuti actually sought first to assimilate tribes peacefully without

Fig. 2.11 Present-day Lake Titicaca on the border of southeastern Peru and northwestern Bolivia. (Haylli / CC BY-SA / https://creativecommons.org/licenses/by-sa/4.0)

bloodshed when at all possible. The Incas did this by sending scouts to assess the tribe, followed by communication with the chief informing him that he could keep his throne if he willingly joined the empire. The Incas often included bribes and most of all made it clear that they would take the land and its people by force if necessary. The tribe could be welcomed as new citizens of the empire or be destroyed by the Incan army (Hemming 1970).

The success of Pachacuti's initial efforts inspired him further as he added several large tribes to his empire, along with assets such as gold and silver (Lee 2000). He carried on a campaign across the Andes that would be continued by Topa Inca and Huayna Capac. Gasparini and Margolies (1980) suggest that the goal for conquest "...was control of certain ecological zones and strong points rather than conversion of every last peasant to the solar cult...." (p. 101). Achievement of such would provide the Incas with security for their civilization as well as locations for fortresses and administrative centers to be used for control of these conquered regions and the imposition of Inca religion and culture upon those living there.

Conquest was not always peaceful, however. Subjugation of the Cañari, for instance, was protracted and quite bloody with great Inca brutality. Willing Cañari loyalty was never secured and thousands of them eagerly sided with the Spaniards against their Inca masters during the ensuing European conquest (Hemming 1970). Others, such as the Wanka, would do so as well (Gasparini and Margolies 1980).

The Incas often would displace new subjects after their conquest. Local gods were removed to Cusco along with their attendants and entire populations were conscripted and dispersed to service distant royal projects (D'Altroy 2002; Hemming and Ranney 1982). Still, Pachacuti and the elites of his administration sought to learn concepts of value from their new citizens while integrating them into their positions in Inca society.

In his legendary conflict with the Chanca, Pachacuti reportedly wore the hide of a Puma into battle. Pachacuti is credited with being the designer of Cusco, a community whose city plan incorporating the confluence of two rivers has been said to resemble the body of a puma, a symbol of military valor that also represented the royal dynasty (Niles 1987 ; Paternosto 1989). Zuidema (1985) counters that this concept is mistaken and that the real meaning of the puma is a metaphor meant to represent the emperor at the head of the Inca state. He asserts that there is no evidence in the chronicles to support the shape of Cusco as a feline. Betanzos (1996 [1576]) writes of Pachacuti having decreed that the far end of Hurin Cusco would be referred to as "Pumachupa, which means the lion's tail" (p. 71), but this still was more figurative rather than part of an intentional design of the city.

The head of Cusco's puma was defined to the northwest by the jagged teeth of Sacsahuaman (Fig. 2.12) and its tail began where two streams met in the southeast. The streets of Cusco weaved throughout the body between (Hemming and Ranney 1982).

As the architect of Inca civilization Pachacuti initiated such magnificent projects as Sacsahuaman, Ollantaytambo, and Macchu Picchu (Gasparini and Margolies 1980). He felt he could and was obliged to improve upon the handiwork of the creator. As the son of the Sun and co-creator of the land, Pachacuti promoted a style of

Fig. 2.12 Zigzag walls of Sacsahuaman. (Author's photograph)

masonry integrated with natural rock formations. Several sites give the appearance that the stone blocks of manmade construction are integral and simply grow from their natural rock foundations (Paternosto 1989). Pachacuti also modified and enhanced stone outcroppings to improve them as huacas, some incorporating astronomical alignments. Celestial orientations found in the huacas and structures created during Pachacuti's reign suggest a strong interest in the solstices.

In his efforts to further develop religion and society, Pachacuti established such innovations as calendrical methods of marking religious ceremonies and rituals, as well as those denoting the proper times for planting and harvesting specific crops (Bauer and Dearborn 1995). These practices were steeped in recurring celestial cycles, especially those involving the solstices and the heliacal rise of the Pleiades.

Topa Inca Yupanqui

Topa Inca (Fig. 2.13) acceded to the throne in 1471 and continued his father's massive campaign of territorial expansion. He also is noted for his construction projects, among them the establishment of his royal residences. Topa Inca claimed the Chinchero valley as the site for his rural estate and set about construction of its palace, courtyard, support buildings, and agricultural terraces. The estate extended to the salt terraces of Maras (Fig. 2.14) and perhaps also included the terraced collapsed basins of Moray (Fig. 2.15). The style of architecture and design suggests a view of nature similar to that of Pachacuti's in which natural rock and landscape

Fig. 2.13 Topa Inca Yupanqui. (Brooklyn Museum / Public domain / https://www.brooklynmuseum.org/opencollection/objects/2307)

features were included in structural forms. Also on the estate are several intricately carved rock huacas (Niles 1999).

While the sites of Pachacuti are steeped with solsticial alignments, his son may have embraced a different philosophy. The many walls and terraces of the Chinchero estate are oriented cardinally, precisely north-south and east-west (Fig. 2.16). This in itself may represent astronomical prowess, as the cardinal direction of south might first have been determined by the shadow plot of a vertical gnomon. North, east, and west could then follow geometrically.

Topa Inca is credited with completing the complex of Sacsahuaman (Fig. 2.12) at the northwest edge of Cusco. He ruled until 1493 (Niles 1999).

Huayna Capac

Huayna Capac (Fig. 2.17) was the son of Topa Inca and Mama Ocllo who, as his father's wife and sister, was both mother and aunt. In this way his parents belonged to a lineage both patrilineal and matrilineal, just as the creator god Wiracocha was said to have founded both patrilineal and matrilineal lineages (Zuidema and Quispe 1973).

Huayna Capac was born at Tomebamba, near modern Cuenca in Ecuador. Niles (1999) states that Pachacuti designated this grandson to succeed Topa Inca to the

Fig. 2.14 Maras Salt Terraces. (Author's photograph)

Fig. 2.15 Terraced Basin at Moray. (Author's photograph)

Fig. 2.16 Walls at Chinchero are oriented east-west and north-south. (Author's photograph)

throne. She says that Huayna Capac remained close with his mother, who made him promise not to leave her to do battle – a pact he honored until her death.

Coya Cusirimay, Huayna Capac's legitimate wife, was his full sister, but did not give him an heir (Niles 1999). He next chose another sister, but she did not receive the blessing of his father's mummy. We will learn more about the importance of such blessings in Sect. 3.7. Ultimately, Huayna Capac took Cibichimpo Rontocay as his principle wife. He is reputed to have had more than 50 others. Huayna Capac fathered scores of children, among them his sons Huascar (Fig. 2.18) and Atahualpa (Fig. 2.19). They were half-brothers and Atahualpa was the elder. When Huayna Capac departed on a protracted campaign to subdue rebellious tribes in the north, he left Huascar in Cusco to govern in his absence and took Atahualpa with him into battle (Niles 1999). This action ultimately would take its toll on the empire.

Huayna Capac built his country estate in the Sacred Valley placing his palace, Q'espiwanka (Figs. 2.20 and 2.21), near the site of the modern village of Urubamba. He, like his grandfather Pachacuti, exhibited interest in the solstices through deliberate astronomical alignments. Huayna Capac additionally built a palace in Cusco and commissioned architecture at Tomebamba, the place of his birth (Niles 1999).

Huayna Capac was skilled in diplomacy and is said to have genuinely cared about his subjects (Niles 1999). He spent a great deal of time to the north, away from Cusco, and considered making Quito a second capital (Hemming 1970).

Fig. 2.17 Huayna Capac. (Guaman Poma, f. 112)

Huayna Capac died without a designated heir in 1527, possibly from the sweeping epidemic of smallpox brought from Europe. His mummy was prepared, adorned, and taken on a litter to Cusco (Betanzos 1996 [1576], p. 185; Guaman Poma de Ayala 2009 [1615], p. 108). Later the mummy was taken to Q'espiwanka where it was hidden from the Spaniards for more than 20 years (Betanzos 1996 [1576], p. 190). A civil war ensued as Huascar, in Cusco, assumed the throne (Sarmiento 2009 [1572]), and Atahualpa, with the Imperial army in Quito, challenged him from the north. The epidemic swept the empire killing thousands, and the civil war claimed many more. The timing of the arrival of a small but determined band of mounted and armored conquistadors couldn't have been better (Hemming 1970; Niles 1999).

Fig. 2.18 Huascar Inca. (Public domain / https://commons.wikimedia.org/wiki/File:Portrait_de_Huascar.jpg)

2.4 Spanish Conquest

Following Huayna Capac's death, Huascar controlled the southern part of the empire and worked to establish himself as his father's successor. Atahualpa controlled the north, however, and had his own designs on the throne. A great battle erupted between them that embroiled and decimated the nation. Atahualpa's followers challenged the legitimacy of Huascar's claim to rule and supporters of Huascar characterized Atahualpa as a usurper of royal power (Hemming 1970). The empire eroded to a considerably weakened state, first due to decimation by smallpox and then by the death toll and chaos of civil war.

Many of the Incas' recently conquered tribes hated their new masters passionately, a factor that served the Spaniards well. These natives found they disliked the Incas more than they did the Europeans, and some seized the opportunity to strike back at their conquerors by aiding the Spanish effort against them. When combined

Fig. 2.19 Atahualpa Inca. (Unknown artist from the Cusco School. Credit: Ethnological Museum of Berlin, Germany)

with plague and civil war, these events softened the battle-hardened empire to the point where the invading conquistadors met with what proved to be insufficient resistance (Hemming 1970).

Fifty-four years after Pachacuti defeated the Chancas, Columbus sailed to the West Indies in 1492. Spain had amassed considerable, but costly, military prowess and sought new fortunes to help fund its expansions. Once word of the riches found by Columbus reached the Old World, the new one would never be the same (Hemming 1970).

Captain Francisco Pizarro (Fig. 2.22) crossed the Panamanian Peninsula and reached the Pacific in 1513. A base for exploring the Pacific shores was founded there and named Panama (Hemming 1970).

In 1519 Hernán Cortés (Fig. 2.23) invaded Mexico and set out to conquer the Aztecs with as few as 500 men and 16 horses. With assistance from some of the Aztecs' subject tribes, Cortés succeeded and subsequently shipped much wealth to his king (Hemming 1970). The promise of untold riches brought many to the Americas, and ultimately Peru, in search of personal fortunes.

Fig. 2.20 Present Q'espiwanka signage. (Author's photograph)

Fig. 2.21 Q'espiwanka. (Author's photograph)

Fig. 2.22 Francisco Pizzaro. (Amable-Paul Coutan / Public domain / https://commons.wikimedia.org/wiki/File:Portrait_of_Francisco_Pizarro.jpg)

The year 1522 saw Pascual de Andagoya explore 200 miles of the Columbian coastline and land at the San Juan River in search of the *Viru*, the name which later gave rise to *Peru*. Pizarro set out in 1524 and 1526 to find his fortune but fell short on both occasions (Cieza de Leon 1998 [1555], pp. 48–51; Hemming 1970).

Before a single European set foot in Peru smallpox, brought to the Americas by livestock, swept the country from the north, killing a great many of the empire's subjects as well as their emperor (Betanzos 1996 [1576], pp. 160–161). Spanish presence in Central America also introduced diseases such as measles and diphtheria, for which the Andeans likewise had little resistance. The effect was devastating as routines normal for maintaining daily life were abandoned and society ground to a halt.

The civil war continued to drain the empire for 5 years until Atahualpa ultimately defeated Huascar and captured Cusco in 1532. Atahualpa's reign would be short-lived, however, as the Spaniards under Pizarro were soon to arrive (Hemming 1970).

Intrigued by the new world riches of Cortés, in 1529 the Spanish Queen authorized Pizarro to conquer Peru and named him as its governor. Taking three

Fig. 2.23 Hernán Cortés. (Real Academia de Bellas Artes de San Fernando / Public domain / https://commons.wikimedia.org/wiki/File:Retrato_de_Hern%C3%A1n_Cort%C3%A9s.jpg)

half-brothers with him he sailed from Panama to Ecuador in 1530 and ultimately reached the northern boundaries of the Inca Empire in 1531. It was there he first found the carnage that had been left by plague and civil war (Hemming 1970).

Atahualpa captured Cusco and placed Huascar under custody, but soon after he was himself imprisoned by Pizarro. The Spaniards later moved into Peru with 62 cavalrymen and 106 soldiers, advancing just as the civil war was coming to an end. Technological superiority of armored and mounted soldiers, prefaced by plague and internal warfare, was largely responsible for Spanish success against massive indigenous forces fighting with slings and spears (Hemming 1970).

Atahualpa sent an envoy to Pizarro and extended him an invitation to meet. The Spaniard agreed but instead plotted to kidnap the Inca, as had been done so successfully by Cortés with the Aztecs in Mexico. The Spanish attack worked and Pizarro held Atahualpa for 8 months before finally executing him in July of 1533. During the interim Atahualpa paid Pizarro a ransom consisting of a great majority of the empire's gold for his release, but this became a bargain that was never honored on the Spanish side. While imprisoned Atahualpa sent an order to have Huascar murdered, so with both of them dead, the Incas were once again left with a vacuum of leadership in a time of great crisis (Sarmiento 2009 [1572], pp. 238–240; Hemming 1970).

Pizarro likely never intended to release Atahualpa, but he collected a great fortune during the Inca's captivity. The Spaniards, concerned with potential rebellion, ultimately garroted the Inca, but only after he first nominally converted to Christianity. King Charles was quite disturbed that Pizarro executed a royal monarch, but several local tribes were pleased to realize an end to Incan oppression (Hemming 1970).

2.5 Inca Resistance

The Spaniards entered Cusco on November 15th, 1533 and took the city without a struggle.Word of Peruvian treasure spread widely throughout Spain, and a rush of fortune seekers descended upon the empire. Soldiers were richly rewarded if they agreed to remain rather than return to Europe (Hemming 1970).

Tupac Huallpa (Fig. 2.24), the son of Huayna Capac and brother of Huascar, was allowed by Pizarro to be coronated immediately following Atahaulpa's execution. The Spaniards were viewed as liberators in Cusco, having rid the empire of Huascar's rival, and Pizarro sought to establish a puppet regime (Betanzos 1996 [1576], pp. 276–277; Hemming 1970). Tupac Huallpa's reign was short-lived, however, as he soon died of illness and once again left Pizarro with the task of installing

Fig. 2.24 Tupac Huallpa. (Unknown artist / Complete History of Peru / https://todalahistoriaperuana.blogspot.com/2018/06/muerte-de-tupac-hualpa.html)

Fig. 2.25 Manco Inca Yupanqui. (Guaman Poma, f. 398)

a controllable head of state. For this he chose another of Huascar's sons, Manco Inca Yupanqui(Betanzos 1996 [1576], pp. 278–279; Fig. 2.25).

Pizarro, concerned about an attack from Quito, now openly embraced the Incas in Cusco. He felt Manco, a legitimate heir, would be a popular choice and saw to it that he was installed as the new leader. The relationship worked at first, but it ultimately was undermined by Spanish greed and arrogance. Spanish treatment of Manco, even as the Incas' leader, eroded over time to where possessions from his home were stolen and ultimately his princess wife was taken by Gonzalo Pizarro for his own. In the wake of such events, Manco evolved from a compliant puppet into a revolutionary leader (Betanzos 1996 [1576], pp. 280–284; Cobo 1983 [1653], pp. 172–177; Hemming 1970).

By 1535 Manco had suffered long enough and realized that his people were faring no better. He decided he could no longer support the Spanish regime and left Cusco under the cover of darkness. Manco soon, however, was captured, imprisoned, and tortured. He was released in 1536 and it was ordered by King Charles that he be given the due of a hereditary monarch. Manco once again fled Cusco and assembled a great army at Calca. Villca Umu, the chief priest of the empire, occupied Sacsahuaman with additional forces. With Manco on the run Pizarro installed another puppet, Cusi-Rimac, in his place (Hemming 1970).

The Incas struggled to kill armored cavalrymen and seldom succeeded. Slings were their best weapons as they laid siege to Spanish occupied Cusco. The thatched roofs of the capital were set on fire in an effort to drive the Spaniards out. Flooded fields, pits, and bolas all were employed to combat the mounted threat. At times it looked as if the Incas might succeed, only to be thwarted once again as the Spanish persevered to victory (Betanzos 1996 [1576], pp. 289–290; Cieza de Leon 1998 [1555], pp. 455–462).

Defeated at Cusco and Sacsahuaman, Manco fell back to Ollantaytambo to make his next stand and secured one of the Incas' few victories against their foes. As Spanish cavalrymen approached the monumental terraced walls of Ollantaytambo (Fig. 2.26), the Incas released great waters and flooded the fields below them. At the same time, attacks were made from the flanks on those caught in the deluge and the Spaniards narrowly were able to flee to safety. The following year, in 1537, the Spanish attacked with better preparation and forced Manco to retreat once again,

Fig. 2.26 Inca Terraces at Ollantaytambo. (Author's photograph)

Fig. 2.27 Diego de Almagro. (Domingo Mesa / Public domain / https://commons.wikimedia.org/wiki/File:Diego_de_Almagro.JPG)

this time to the fabled Vilcabamba. Much to the consternation of Pizarro, as long as the Inca lived so did the empire (Betanzos 1996 [1576], pp. 290–291).

Diego from Almagro (Fig. 2.27), an ambitious rival of Pizarro, seized Cusco in 1537 leaving the Spaniards with a civil war of their own. Paullu Inca (Fig. 2.28) assisted Almagro, who later rewarded him with his brother Manco's title. The battle between Pizarro and Almagro continued into 1538 when Francisco's brother, Hernando Cortez, recaptured Cusco and had Almagro executed (Hemming 1970).

The Inca rebellion continued in 1539, and while deep in the forests of the Amazon, Manco's forces established Vilcabamba (Fig. 2.29) as their new capital. Vilcabamba grew to be a significant settlement and served to preserve Inca culture and religion. Pizarro and Manco killed each other's men in skirmishes, but Vilcabamba remained secure, with Incas in Cusco comforted in the knowledge that their emperor was still alive (Hemming 1970).

Paullu thrived under the Spaniards and was later baptized as a Christian. It was Paullu who ultimately gave up the mummy of his father, Huayna Capac, from its hiding place at Q'espiwanka. Sarmiento de Gamboa (2009 [1572], pp. 217) relates that it was Polo de Ondegardo who found the emperor. The Incas in Cusco did not regard Paullu as their emperor, however, and instead looked for guidance from Manco; many defected to join him. Francisco Pizarro was murdered by sympathizers of Almagro in 1541, who then quickly moved to install Almagro's son as

Fig. 2.28 Paullu Inca. (Just Cala Sahuaraura / Public domain / https://commons.wikimedia.org/wiki/File:Paullu_1838.jpg)

Governor of Peru. The assassins fled and found refuge with Manco in the jungle. Almagro's son was put to death in 1542 (Cobo 1983 [1653]; Hemming 1970).

The Spaniards showed little interest in Inca astronomy and at the time of the conquest were still influenced by the geocentric teachings of Claudius Ptolemy. Copernicus' heliocentric theory of the universe was introduced in Europe in 1543 and received little acceptance for decades. The conquistadors were preoccupied with wealth and power, while the interests of the Catholic priests largely centered upon the extirpation of indigenous Sun-worship idolatries.

Fig. 2.29 Ruins of Vilcabamba. (I, AgainErick / CC BY-SA / http://creativecommons.org/licenses/by-sa/3.0/)

Manco welcomed the Almagrists, but they grew weary of their exile in the rainforest. In 1544 they reasoned that they might be able to return to Cusco by ridding the Spaniards of Manco and thereby stabbed him to death. None of them escaped, however, as all were quickly hunted down while they fled and put to gruesome deaths as well (Cobo 1983 [1653], pp. 175–177; Hemming 1970).

Manco lived long enough to learn of his assassins' demises and also to name his son, Sayri Tupac (Fig. 2.30), as his successor (Betanzos 1996 [1576], pp. 297–299; Cobo 1983 [1653], pp. 175–177). Sayri Tupac was 5 years old at the time and ruled peacefully for 13 years through his regents and advisors. He was influenced by his uncle in Cusco, Paullu Inca, and significantly reduced hostilities toward the Spanish. Sayri was enticed by Paullu to leave exile for Cusco. In 1557 he was given a full pardon and a large estate in Yucay, where he and his wife, Maria Cusi Huarcay, both having been baptized, lived until his death in 1561 (Farrington 1995). This left the Spaniards no living Incan heir in Cusco (Hemming 1970). Niles (1999) says it was widely believed Sayri Tupac was poisoned.

Meanwhile Titu Cusi Yupanqui (Fig. 2.31) usurped Tupac Amaru as Inca upon Sayri Tupac's departure from Vilcabamba. Titu Cusi was a very capable ruler and was well supported by the empire's military commanders. The Spaniards tried to entice Titu Cusi to leave Vilcabamba for Cusco. He played them masterfully for years, but never left the sanctity of his jungle refuge (Hemming 1970).

Fig. 2.30 Sayri Tupac. (Justo Sahuaraura / Public domain / https://commons.wikimedia.org/wiki/File:SAHUARAURA_p115_-_SAYRI_TTUPAC.jpg)

By the 1560s, as Spanish priests learned Quechua, they became aware of a great resurgence of indigenous Sun-worship and idolatry, in part inspired by perpetuation of Inca culture in Vilcabamba. Eradication of everything associated with these pagan beliefs became of top concern for the Church, and renewed effort was put into the campaign initiated for that purpose (Hemming 1970).

Titus Cusi became so adept at giving the Spaniards just enough hope for a peaceful conclusion that he was able to forestall an invasion for years. In 1568 he learned the catechism and was baptized by a priest who had braved the jungle. He was christened Don Diego de Castro Titu Cusi Yupanqui, but came to resent some of the lifestyle constraints the friars tried to impose (Cobo 1983 [1653], p. 181; Hemming 1970).

Francisco de Toledo (Fig. 2.32) was appointed as the fifth Viceroy of Peru in 1569. He was said to be a great administrator and he instituted tenets of government that functioned well for generations. Conversely, Toledo legalized the mita labor system with the result being a great number of deaths and a drastic decrease in population. He resented the former Inca monarchy and argued that the empire was no more legitimate in its rule of Peru than were the Spanish – both were conquerors

Fig. 2.31 Titu Cusi Yupanqui. (Titu Cusi Wikipedia CC BY-SA / https://alchetron.com/Titu-Cusi#titu-cusi-174737fb-6053-44e1-8acd-d040379a357-resize-750.png)

of indigenous tribes. He maintained that the Incas were decadent in the eyes of God and that the Inca nation in Vilcabamba could not be allowed to continue to exist (Hemming 1970).

In May of 1571 Titu Cusi became ill and died, but the Incas kept his passing a secret from Cusco. Some thought he had been poisoned and Martin Pando, Titu Cusi's mestizo advisor, was promptly executed. The other person in Vilcabamba with Spanish ties, Friar Diego Ortiz, was stripped and bound, but only killed later on command of the new Inca (Hemming, 1970).

Titu Cusi's captains decided that his brother Tupac Amarú (Fig. 2.33) should lead them instead of his son, Quispe Titu. Tupac Amarú rejected Christianity in favor of traditional worship. He closed Vilcabamba to all from the outside and ordered obliteration of anything Christian. Peaceful coexistence with the Spaniards came to an end (Hemming 1970, p. 418).

Toledo did not know of Titus Cusi's death and continued efforts of diplomacy. He dispatched an emissary who was killed to preserve the secret. By early 1572 Toledo was sure he needed to conquer Vilcabamba and sought the King's permission to attack what was viewed as a sovereign monarchy. War was declared on April 14th and Toledo prepared for the invasion which ensued soon after. On June 24th the

Fig. 2.32 Francisco de Toledo. (Credit: National Museum of Archaeology, Anthropology, and History of Peru)

Spaniards marched into Vilcabamba, but found it deserted and burned, with the Incas having retreated further into the jungle (Hemming 1970).

Quispe Titu and his wife were captured in the rainforest 6 days later. The Manari's helped the Spanish locate Tupac Amaru, who was captured along with his pregnant wife some 275 kilometers from Vilcabamba. Tupac surrendered peacefully and was promised safe passage and fair treatment. Upon his arrival in Cusco, however, Toledo had him immediately tried, convicted, and sentenced to death (Hemming 1970).

Like several of his predecessors, Tupac Amaru was instructed in Christianity and baptized before his execution. At the gallows, this final Inca gave a compelling speech to a massive crowd of his subjects. He denounced paganism and extolled Christianity. When finished he received his blessing and was then killed (Hemming

Fig. 2.33 Tupac Amarú. (Source unknown / Public domain / https://commons.wikimedia.org/wiki/File:TupacamaruI.JPG)

1970; Lee 2000). The lineage of Inca kings had come to an end. After a struggle lasting for 40 years, the conquest was finally complete and the Spaniards were in total command of Peru.

2.6 The Catholic Purge

The Catholic Church saw as its mission to convert all it encountered to Christianity (Hemming 1970). Natives in foreign lands, previously unexposed to Church teachings, were considered of prime importance for religious education and to be saved from their pagan ways. The Vatican often failed to acknowledge that many native cultures had deeply rooted belief systems that would be hard to eliminate. This certainly was the case with the Incas.

Spain sought Vatican permission to conquer the Americas and was granted this by Pope Alexander VI, as long as they converted all they dominated to Christianity. Chapels, cathedrals, and monasteries were built in the major centers of Peru, but progress was limited at first due to the insufficient number of priests available to do the work. The Spaniards were somewhat successful in suppressing public practice of the Incas' religion, but found certain fundamental elements hard to eliminate (Hemming 1970).

The Spaniards looted and destroyed the most important shrines, such as the Temple of Pachacamac and the Coricancha of Cusco (Fig. 2.34), soon after their invasion of the Inca homeland. In 1539 the Spanish began a ruthless campaign against the indigenous religion and set about to systematically destroy as many huacas as possible (Arriaga 1968 [1621]; Bauer 1998). Attendants and worshippers of known huacas were prosecuted, tortured, and/or put to death, the foundations of the shrines were dug out, the object of worship was destroyed, anything flammable was burned, and finally a cross often was built over the ravaged space. As part of this destructive campaign, the names and locations of these huacas were recorded so that they could be checked in the future to make certain no religious activity continued. Sometimes the only extant evidence, this information has played a valuable role in

Fig. 2.34 The author and research assistant son, Steven, at the Coricancha; the Spaniards built atop the original Inca site. (Author's photograph)

huaca research. Other huacas, essentially those carved in stone, could not be obliterated and remain to this day. The Spanish lacked the ability to destroy them and some have been preserved with remarkable integrity.

In 1567 an ecclesiastical council was conducted in Lima with its aim to eradicate pagan rites. Hemming (1970) tells us that priests were instructed to abolish superstitions, ceremonies, arrest witch doctors, and destroy any shrines or talismans. With this action the Church began an aggressive effort to erase all elements of indigenous religion. Father Pablo Joseph de Arriaga described in detail the process of idolatry extirpation followed in Peru. He told of the indigenous worship and how it should be eradicated. Arriaga (1968 [1621]) said that the "moveable huacas" once discovered were taken away and burned. Huacas that could not be removed, such as "high hills and mountains and huge stones" could still be worshipped. Of these he stated "…we must try to root them out of their hearts, showing them truth and disabusing them of error" (pp. 24–25). The Spanish succeeded in destroying the most important shrines of Cusco, but the priests remained unable to obliterate rock and the life forces believed by the Incas to be contained within.

Incan temples were either destroyed or their stone bases were used as the foundations for structures such as Catholic chapels (Fig. 2.34). Many ruins survived, however, in remote areas. The extirpation of idolatries consumed members of the clergy for decades (Bauer 1998; Gasparini and Margolies 1980; Hemming 1970).

2.7 Summary

To comprehend fully the intrinsic celestial alignments of Inca huacas, the shrines must be examined not just astronomically, but also holistically with regard to their function in society and cosmology. An understanding of the history of Inca culture and what influenced its cosmological and religious beliefs is required before it can be determined why these astronomical orientations grew to be of such great importance.

The Inca Empire was relatively short-lived, but Inca culture was built upon a long line of preceding societies. This is important for Incan astronomy because certain celestial knowledge had to be obtained cumulatively through observations of repetitive cycles made over very long periods of time. Astronomy was a major part of Inca cosmology, religion, and agriculture, and its significance as such is evident in many of the shrines and structures left by the emperors Pachacuti, Topa Inca, and Huayna Capac.

Our knowledge of Inca astronomical practices suffers from the lack of a written language and the desire of the Catholic Church to eradicate anything viewed as being related to indigenous religion. This extirpation of idolatries continued for decades and destroyed countless shrines. The Church accelerated its efforts during the latter part of the sixteenth century when a religious resurgence was discovered, partly inspired by the protracted survival of Inca culture in Vilcabamba.

Records of huacas were maintained by the Church after the shrines were destroyed, and Spanish chroniclers recorded information from conversations held with local informants. These do not give true contextual perspective to Inca astronomy, however, as the writings were from a European point of view at a time when even the teachings of Copernicus were new. The Spaniards failed to realize that the Incas saw the cosmos in their own unique and sophisticated way.

As part of the extirpation effort, the Spaniards sought to obliterate anything related to worship of the Sun or the Earth. Eradicating all evidence of solar pillars or other constructs designed to monitor the progress of the Sun was of prime importance as the Incas were a Sun-worshipping society (Bauer and Dearborn 1995). The efforts of the Catholics were quite effective and Incan horizon astronomy quickly faded from use.

Certain carved rock huacas and temple ruins survived the purge and provide us with a glimpse of Inca astronomy through examination of its celestial orientations. The remote solar pillars at Q'espiwanka remain and certain isolated sites, such as Machu Picchu, were never discovered by the Spanish. Untouched by the priests, they are of prime importance for research.

Chapter 3
Cultural Context

Fig. 3.1 Girl with Cria (baby llama) in Cusco. (Watercolor by Jessica Gullberg)

© Springer Nature Switzerland AG 2020
S. R. Gullberg, *Astronomy of the Inca Empire*, Historical & Cultural Astronomy,
https://doi.org/10.1007/978-3-030-48366-1_3

Astronomy was an integral part of Andean mythology and creation, and it was at the very heart of the Incas' religion and agriculture. To understand Inca astronomy, it must be placed into context in the greater society. Why was it important, why was it used, and how was it used? This chapter will discuss many aspects of Incan culture to show its association with astronomy as well as other factors that influenced Incan thought and reasoning.

3.1 Religion

The Incas proclaimed themselves to be the children of the Sun. They worshipped it and viewed their emperor as being the Sun's direct descendant.

The Incas benefitted from existing Andean astronomical knowledge and beliefs and made solar worship the official religion of their empire. Pachacuti imposed it across the realm, maintaining that he was the son of the Sun and his wife the daughter of the Moon. The Incas venerated the Sun, the Inca, and his predecessors. Their religion was tied closely to nature with the prosperity of the world at the whim of supernatural forces found within mountains, caves, and streams, as well as in huacas and celestial objects such as the Moon, stars, rainbows, and thunder (Hemming 1970; Hemming and Ranney 1982). Conquered tribes were made to accept state religious beliefs but were allowed continued worship of their lesser gods.

The ruling Inca was the central figure in solar worship, supporting the assertion that he was the descendant of the Sun. The emperor was the intermediary between the heavens and the populace of the realm (Bauer and Dearborn 1995; Bauer and Stanish 2001). This deification was necessary to solidify the ruling Inca's legitimacy, to justify his absolute authority, and also establish that of the state (Hemming 1970).

The Incas associated the Moon with *Coya*, the queen of the ruling Inca (Bauer and Stanish 2001). The Moon served as both the wife and the sister of the Sun, a relationship that was also promoted for the empire's ruling couple (Zuidema and Quispe 1973). In Inca culture, the Moon was feminine and the Inca's Coya was its daughter. Women worshipped the Moon and made offerings to it during eclipses and when giving birth. Descriptions of Inca lunar worship are relatively rare, perhaps because of its feminine role in society (Bauer and Stanish 2001).

State worship took place at temples and huacas, as well as in pilgrimage centers designed to promote the authority of the ruling elite. Private worship proliferated with the veneration of deities located in niches built into walls of the home (Fig. 3.2). Individuals led lives filled with rigid rituals they felt would enhance their personal wellbeing (Hemming 1970).

Pachacuti established his religion as a state institution and organized a priesthood to attend to its temples and huacas. He constructed many temples as he expanded his empire and decreed solar worship for everyone. The three most significant sites of worship became the Temple of Pachacamac (Fig. 3.3), the Coricancha (Figs. 2.35 and 6.3), and the Islands of the Sun and Moon in Lake Titicaca (Bauer and Stanish 2001; Fig. 2.11).

Fig. 3.2 Example of niches in a wall at Machu Picchu. (Author's photograph)

Fig. 3.3 Pachacamac. (Charles Gadbois / CC BY / https://creativecommons.org/licenses/by/3.0)

3.2 Cosmology

The Incas believed the world was created by their god Wiracocha, or Viracocha, at Lake Titicaca (Fig. 3.4). Wiracocha was thought to be the father of the Sun and the Moon. He was both male and female, which enabled him to be the founder of both patrilineal and matrilineal descent lineages (Zuidema and Quispe 1973). Wiracocha first made people of stone and then made the Sun, stars, and Moon. He gave the stones life as they appeared from caves, rocks, and springs (Paternosto 1989). The Incas worshipped many deities in a hierarchal pantheon.

A common Inca creation myth related that Manco Capac (Fig. 2.7), with his three brothers and four sisters, left Lake Titicaca in a migration beneath the Earth and emerged from a cave south of Cusco (Hemming and Ranney 1982). This is important

Fig. 3.4 Image representing the Inca god Wirachoca. (Unidentified painter / Public domain / https://commons.wikimedia.org/wiki/File:Viracocha.jpg)

as caves and other natural features of the Earth became focal points of Inca veneration. They viewed caves and rock outcroppings as connections with their underworld.

Inca cosmology begins with Lake Titicaca, a large lake at an altitude of 3810 meters above seal level (masl) on the border between present-day Peru and Bolivia (Fig. 2.10). The temperature-retentive waters create a microclimate very conducive for growing at that altitude, a factor that has attracted humans to the region since at least 2000 BCE. Earliest occupation may date prior to 6000 BCE and perhaps long before (Bauer and Stanish 2001).

According to Bauer and Stanish (2001), the area surrounding Lake Titicaca took on more than local importance during the Tiwanaku period of 400 CE–1100 CE. It was perhaps then, long before the Incas, that the lake and its islands first drew pilgrims for reasons of cosmological significance.

Pachacuti conquered the Lake Titicaca region, sent thousands of mitmaes to occupy it, and also was alert for local practices that might be beneficial to his empire. Legend claimed the Island of the Sun and the Island of the Moon to be the points of origin for the Sun, the Moon, and the entire Inca tribe. As such, Pachacuti was quick to capitalize on these shrines and quickly incorporated both into cultural practice regarding Inca origins (Dearborn et al. 1998).

Pachacuti constructed a temple of the Sun and shrine to the Moon on the islands and, as the Incas' cosmological points of origin, they were instituted as state-sponsored ritual pilgrimage centers (Bauer and Stanish 2001).

The Incas believed that the Sun first rose from a sacred rock called Titicaca on the north end of the Island of the Sun, thus annual rituals were orchestrated there for both the June and December solstices. Non-elites were unable to view solar alignments at the sacred rock itself and were instead required to observe from a platform at a distance (Bauer and Stanish 2001). Ceremonial class separation seemed to be a recurring theme, with potential examples observed during the author's field research at Urubamba and Tipon.

The cosmos of the Incas existed in three distinct worlds – that of Ucu Pacha, the underworld, Kay Pacha, the here and now, and Hanan Pacha, the world above (Urton 1981a). There are many extant examples of symbolic sets of three stairs representing transition between the three worlds of the Incas' being (Fig. 3.5). Caves figure prominently in Inca origin myths and were thought also to be chthonic connectors to the underworld (Van de Guche 1990).

3.3 Sacred Landscape

The Incas venerated natural features such as mountains, outcroppings, caves, springs, and rivers, all believed to be endowed with sacred powers. They gave high reverence to mountains and the great entities within them. Sacred mountains were prominent on the horizons of Cusco and Machu Picchu and the Incas' great respect for the Earth was no better displayed than in their worship of these majestic snow peaks. Quechua populations today view mountains either as powerful deities

Fig. 3.5 Symbolic steps in the cave within Kenko Grande. (Author's photograph)

themselves or the residences of deities. They are worshipped as ancestors, sources of water and weather, and in the case of Ausengate, the father of alpacas and llamas. Mountains were often venerated as the most important of deities throughout the empire (Reinhard 1985, p. 306; Fig. 3.6).

Similar to sacred mountains, many rock outcrops were also understood to be hierophanies, or manifestations of the sacred. Pachacuti apparently believed that he could improve upon these revered stones and, as the son of the Sun and co-creator of the land, he could modify and enhance the work of the creator. The carved huacas are bedrock features with their roots in the Earth, an important aspect of the symbolism involving the three worlds. They also seem to be laid out across the landscape in meaningful patterns. A limited number of motifs were used in the shaping of huacas suggesting that the carvings were not a form of mindless or inventive graffiti, but perhaps elements in a symbolic language with cosmological significance (Fig. 3.7).

3.4 Camay

The Incas believed that all things had a point of vitalization, or *camac*, and this for a group of humans was their huaca of origin. Camay was *a concept of specific essence and force, 'to charge with being, to infuse with species power'* and camac

Fig. 3.6 Snow peak viewed from Llactapata. (Author's photograph)

Fig. 3.7 Kim Malville examining carvings in the Chinkana outcropping at Chinchero. (Author's photograph)

Fig. 3.8 Channel and ducts for flowing water measured by Carlos Aranibar at the River Intihuatana in the Urubamba/Vilcanota River gorge between Machu Picchu and Llactapata. (Author's photograph)

was one "who charges the world with being" (Salomon and Urioste 1991). The camac for llamas was a constellation in the shape of a llama that, upon descending to Earth, was responsible for giving a general vitality that allowed for all terrestrial llamas to thrive. Camay was thought to give form and force and to animate (Salomon and Urioste 1991; Taylor 1974) in a continual process that brings something into being by energizing matter.

Running water was understood to be an energizing and animating life force in Andean cultures and was also associated as an agent of *camay*. In the cosmology described in the *Huarochirí Manuscript*, life is born from the embrace of feminine Earth by masculine water, homologous to the growth of plants from soil when moistened by water (Salomon and Urioste 1991). The circulation of running water and the pouring of offertory liquids could animate certain inanimate objects to become huacas, which were understood to be sentient beings with extraordinary and superhuman powers. Running water was found near many huacas (Fig. 3.8).

Water empowered huacas through a life-energizing force that could be used to provide sentience to the inanimate or renew power in the living (Salomon and Urioste 1991). The world's water cycled through the heavens and Earth in its

Fig. 3.9 Zigzagged channel at Chinchero. (Author's photograph)

journey down the Vilcanota with return via the Milky Way (Urton 1981a). Yet cosmology viewed the Milky Way as a river flowing across the night sky in a very literal sense. They saw earthly waters as being drawn into the heavens and then later returned to Earth following a celestial rejuvenation. The Earth was thought to float in a cosmic ocean (Urton 1981a). When the celestial river's orientation was such that it dipped into that ocean, waters were drawn into the sky. "The Milky Way is therefore an integral part of the continuing recycling of water throughout the Quechua universe" (Urton 1981a, p. 60).

Salomon and Urioste (1991) suggest that huacas were understood to be living, energized beings brought to sentience by the Earth's waters. A common motif found at several huacas is that of straight or zigzagged channels constructed for the purpose of guiding ceremonial fluids (Paternosto 1989; Figs. 3.9, 7.9, and 8.8). A flow of energy was stimulated by the pouring of liquid offerings into these channels. Reinhard (2002) suggests the importance of an Incan water cult at Machu Picchu. One example of this is found in the stone-lined channel that leads from the double-jambed door of the Llactapata Sun Temple that points across the River Intihuatana to Machu Picchu's Sacred Plaza and beyond to the approximate horizon location of the sunrise at June solstice (Fig. 9.48).

Life was given to inert matter through the action of the cosmological lifeblood of flowing water cycling from the ocean to the sky and back to the land. The powers of camay were great as objects, stones, and even places could be animated through running water or offertory liquids. Snow peaks were especially revered because the Incas recognized them as their immediate source of water in this cycle (Salomon and Urioste 1991).

While many huacas were animated by the camay of water, this concept was not exclusive to fluids. All things were said to have vitalizing entities, or *camac*. A llama-shaped constellation infused earthly llamas and the camac for humans was their huaca of origin. Therefore, energized huacas were responsible for bringing life to others (Salomon and Urioste 1991).

The Incas also believed that rock could be empowered and energized by elaborate carving (Paternosto 1989; Fig. 3.7) as displayed by those huacas that are intricately sculpted outcroppings. Carved huacas were given life, however, through the circulation of the Earth's life-force, its running waters, and all of the astronomical shrines in this book were found to have such a hydraulic source nearby.

3.5 Intihuatanas

Hiram Bingham, of Yale University, is credited with the scientific discovery of Machu Picchu in 1911, and through *National Geographic* he introduced it to the world. During his research, Bingham identified a carved stone adjacent to the Urubamba River (Fig. 3.10; Sect. 9.3) as an intihuatana, perhaps guided to such a conclusion by local informants. He described his understanding of intihuatanas in the following passage:

> You means "sun" and huatana is "a place where animals are tied." The intihuatana would seem to be "the place to which the sun was tied," so that it could not escape. A primitive folk so extremely dependent on the kindly behavior of the sun as were the Peruvian highlanders must have been in terror each year, as the shadows lengthened and the sun went farther and farther north, that he would never return but would leave them to perish of cold and hunger. Hence it seems likely that these short stone posts represented the post to which a mystical rope was tied by the priests to prevent the sun from going too far away and getting lost. (Bingham 1930, p. 52)

There are differing interpretations of intihuatanas and "hitching post of the Sun" may be more of a modern myth inspired by local villagers. The term "intihuatana" only began to appear during the nineteenth century. Even should Bingham's explanation be applicable in certain cases, other suggestions that the carvings were devised as astronomical sighting devices are not supported. Although often identified by tour guides as a calendrical shadow-casting gnomon, there is really no evidence that the Intihuatana stone of Machu Picchu was used for observing and establishing dates of the solstices or even the zenith Sun (Fig. 3.11).

Intihuatanas are also found at Pisac and Tipon. In Pisac the intihuatana is a large, partially carved rock in the temple group that is enclosed by a semicircular masonry

Fig. 3.10 The River Intihuatana. (Author's photograph)

Fig. 3.11 Intihuatana at Machu Picchu. (Author's photograph)

Fig. 3.12 Intihuatana at Pisac. (Author's photograph)

wall adjoining a straight masonry wall (Fig. 3.12). It displays a stone gnomon on its flat upper surface within the walled enclosure. Squier (1878) was informed by the Governor of Pisac that this gnomon had been once clad with a bronze sheath. The primary rock extends beneath and beyond the wall of the structure where a second gnomon is located.

The Intihuatana of Tipon (Fig. 3.13) exhibits a different style in that the in situ rock remains unimproved although a platform has been built around it. The June solstice sunset can be viewed over the Intihuatana from a vantage point on a mesa to the east of the site's extensive terrace system.

3.6 Sacred Animals

Many animals, both living and symbolic, were revered and worshipped in Andean culture. Of the highest order, the Incas venerated the condor, puma, and serpent with regard to their sacred correlations with the three cosmological worlds of the sky, Earth, and underworld.

Urton (1981a) relates Polo de Ondegardo to have said that "in general, [the Incas] believed that all the animals and birds on the earth had their likeness in the sky in

Fig. 3.13 Intihuatana at Tipon. (Watercolor by Jessica Gullberg)

whose responsibility was their procreation and augmentation" (p. 169). The dark cloud constellations recognized in the Milky Way are replete with examples of animals such as the serpent, the fox, the llama, the toad, and the tinamou, each bearing supernatural influence over its terrestrial counterparts (Urton 1981a; Fig. 3.14 and Fig. 6.15).

3.7 Ancestors

The Peruvian archaeologist, Julio Tello, recognized that ancestor veneration has been one of the major and enduring features of Andean civilizations (DeLeonardis and Lau 2004). Huacas appear to be major elements in Andean cosmology extending back to 1000 or 2000 BCE and often were shrines to ancestors who, it was believed, could influence the living. Feeding of huacas was a major motivation for communication with ancestor-gods and for sacrifices (Benson and Cook 2001). Mummies of Inca emperors were considered as deified ancestors (Zuidema 1983).

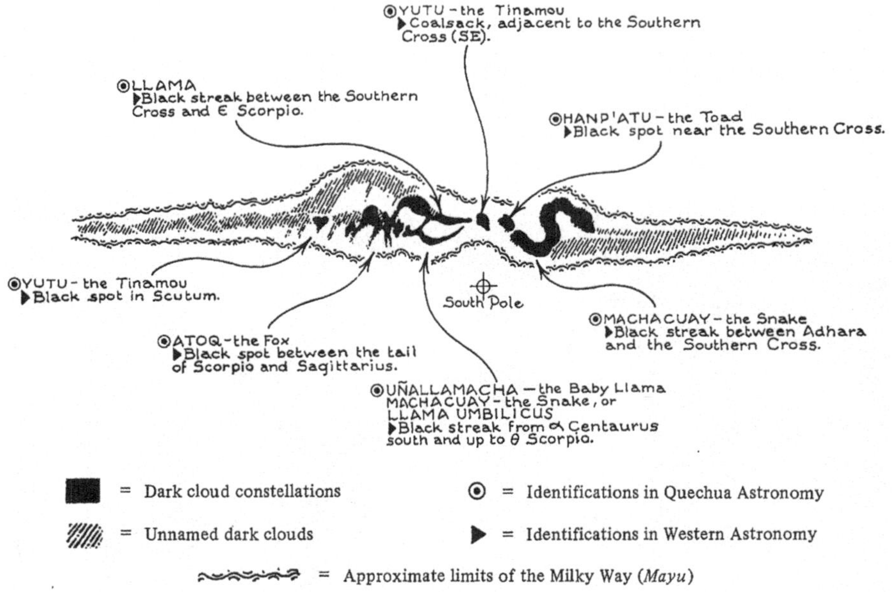

Fig. 3.14 Inca dark constellations in the Milky Way. (From Urton 1981a)

They were carried in processions or placed on platforms, carved steps, and altars. Shamanic communication with the supernatural world of the ancestors and movement between the three cosmological worlds were intertwined with ancestor worship (Eliade 1972; Fig. 3.15). Ancestor worship included the Sun and Moon as well as fundamental themes in origin mythologies such as rocks, caves, and water. Huacas were often places where the ancestors could be called upon for assistance in agriculture, warfare, health, and fertility.

Mummies of ruling Incas played a significant role in Inca culture and worship of them was an integral part of state religion. These preserved royal bodies were treated as if still alive (Niles 1987). When a ruling Inca died his *panaca*, or royal descent group, continued to manage his property and wealth as if he were still living. Each new Inca was required to build a palace and establish his own estate and riches. The mummy continued to occupy his palace, was clothed and "fed," and was consulted on matters of significance. The Inca's descendants remained responsible for his mummy and possessions and would help the mummy participate in state ceremonies (Hemming and Ranney 1982). Mummies sometimes were called upon to visit other mummies, or even the living (Niles 1999). All mummies of the ruling Incas were paraded at a new coronation, an exercise that emphatically displayed the dynastic lineage of the empire.

Fig. 3.15 Drawing of Inca emperor procession. (Guaman Poma, f. 289)

3.8 Social Issues

Royal Marriage

The ruling Inca had many wives, but only one was primary. It was prescribed he marry his own sister because he and she would then belong to a lineage both patrilineal and matrilineal and therefore the emperor would lead a social hierarchy similar to the creator Wiracocha's cosmological hierarchy (Zuidema and Quispe 1973). They were married on the day of his accession as ruler and the children she bore also became primary, the eldest son heir to the empire (Zuidema 1964). Primary children were born by the primary wife and subsidiary children by the Inca's subsidiary wives. In the event of the death of a primary wife, the Inca was required to marry another woman qualified as such.

Moieties, Suyus, Panacas, and Ayllus

The Incas divided Cusco and many other locations into upper and lower halves, or *moieties*. These divisions were both geographical and social. The upslope half of Cusco was *hanan* (upper), while the lower half was *hurin* (lower). Hanan Cusco held higher status than did Hurin Cusco, which was closer to the non-Inca population (Zuidema 1983). There were also parallel social halves with the first five royal descent groups assigned to Hurin Cusco and the next five to Hanan Cusco. The upper and lower moieties were considered equivalent (Zuidema 1964). Gasparini and Margolies (1980) felt that the concept of such upper and lower divisions may have originated in the mountains from the need for coordination between different ecological zones.

Both Cusco and the empire were divided into quarters, or *suyus*. These were called *Chinchaysuyu, Antisuyu, Collasuyu,* and *Cuntisuyu* (Fig. 3.16). The some of Cusco were organized within these divisions.

A *panaca* was a patrilineal royal descent group established by a newly installed Inca. The panaca supported the emperor and was typically led by his second son as the first son would be required to leave the panaca once he succeeded his father. As successor to the throne he inherited the empire, but not any of his father's wealth or possessions. A panaca took on additional importance upon the death of its patriarch as it then became responsible for the care and ceremonial functions of his mummy (Niles 1999).

An *ayllu* was a non-royal extended kinship group from the same patrilineal ancestor that provided structure for the regulation of marriage and inheritance. It was the basic social unit beyond immediate relatives (Niles 1987). It has been said that there were ten ayllus, five in Hanan Cusco and five in Hurin Cusco, which were paired in the same ceque groups as the royal panacas(Sarmiento 2009 [1572], pp. 73–75; Zuidema 1964).

Panacas and ayllus were assigned to specific some and were responsible for their care and that of their associated huacas. Ceques were normally divided into groups of three, *collana, payan, and cayao*, the payan ceques associated with panacas and the cayao ceques with ayllus (Zuidema 1964).

Money

The Incas lived in a society without personal property or money. The state provided for them and determined the labor they would contribute for the common good. Following the Spanish conquest it was difficult for many Inca citizens to grasp the concept of earning money to spend for their needs (Hemming 1970).

Writing and Remembrance

The Incas never developed a system of writing and they therefore have no traditional recorded history. All dates and events prior to the Spanish conquest are approximate from verbal accounts. Early Spanish chroniclers attempted to relate culture and history through interviews of Inca citizens, but in many ways failed in their task due to cultural preconceptions that left them unable to truly comprehend Inca society (Lee 2000; Niles 1987). Art, architecture, and carvings remain as a primary sources of Inca history.

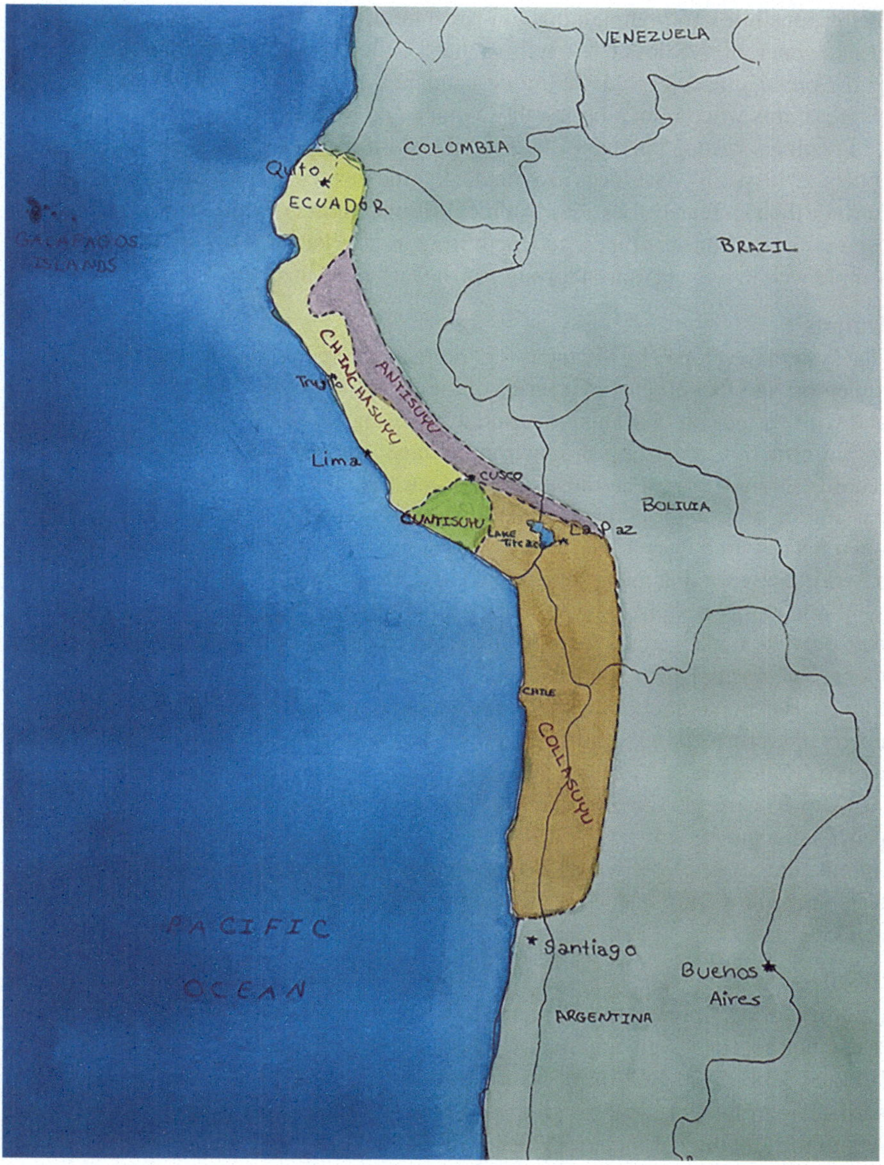

Fig. 3.16 The four suyus of Tiwantinsuyu. (Watercolor map by Jessica Gullberg with suyu lines from Staller, 2008)

Other means for transmitting thoughts and ideas evolved in tapestries and weavings that included the knotted quipu. These artistic substitutes contributed to the hindrance of the invention of an alphabet (Paternosto 1989).

The Incas considered it of primary importance to preserve royal history. Oral legends were recited during ritual festivals, telling tales of dynastic greatness. Even the responsibility for maintaining non-royal family history was passed from generation to generation. Niles (1999) tells us that "…the deeds of an ancestor were related to the prestige accorded his living descendants…" (p. 24). Relatives were acutely aware of this when remembering those that came before.

Lacking a written language "Inca royal histories were works of oral literature…" (Niles 1999, p. 28). The Incas also made graphic history of their battles in conquest, turning their defeated foes into trophies. Human skin covered drums, heads were used as flasks, bones as pipes, and teeth worn as necklaces. Many artifacts and much temple wealth were acquired through such warfare (Niles 1999).

Quipus

The system developed by the Incas for "permanent" recording was that of the *quipu*, a memory aid that employed a series of knotted strings and cords in a certain order and with various colors. This substitute for writing was used as a means of preserving information regarding production, storage, distribution, census data, and tax records (Niles 1987 ; Paternosto 1989). One function of quipus was to aid in the passing of oral histories from generation to generation (Sherbondy 1992). Based upon his research Gary Urton finds that quipus were a system of writing, or at least a hybrid between that and a mnemonic aid (Urton 2003, 2011).

While numbers were known to many, a limitation of the quipu was the likelihood that only its maker could fully interpret much of the other information recorded there. The Incas had a special class of civil servants, *quipucamayos*, who were trained as experts in recording and interpreting quipu data (Fig. 3.17). The quipucamayos therefore had great influence over the content and meaning of Inca official records (Niles 1987).

Quipus were also used to record Inca calendrical information and may have provided a mnemonic aid for family oral histories. Ceques were recorded on quipus, one example of which is the obvious conceptual correlation with the pattern of ceques and huacas surrounding Cusco (Niles 1987; Paternosto 1989; Zuidema 1977).

Textiles

Another form of preserving elements of a society's history is in that of its textiles. Weaving preceded even ceramics as an art form in the Andes. Textiles dating as far back as 3000–2500 BCE have been recovered that display the condor, puma, and serpent – prominent symbols still used four millennia later by the Incas. Textiles record many thoughts such as evidence of the gods and other beings that influenced or concerned these ancient cultures. The Incas continued the use of ceremonial textiles and the art form still exists in the Andes today (Paternosto 1989; Fig. 3.18).

3.9 Organization

All of the important government positions were held by members of the Inca royal family. Other nobility served in supportive administration roles (Hemming 1970).

Fig. 3.17 A quipu and specialist for the calendar. (Guaman Poma, f. 360)

The Incas divided their world into four quarters and from these their empire evolved. They thought of Cusco as the navel of the world and the center of their universe and it served as both a government and ceremonial center (Niles 1987). Cusco became the focal point of Inca religion.

Homes largely were made of masonry with thatched roofs and were generally built on hillsides to save more valuable land for crops. Community sites were selected so as to be a short distance from the fields where their residents would work (Hemming and Ranney 1982; Fig. 3.19). Streets were made of stone with central water channels and constructed in grids. Architecture represented a form of status and residents of similar stature lived in similarly constructed homes (Niles 1987; Fig. 3.20).

Fig. 3.18 Andean textile. (Credit: Chancay sleeved tunic with flying condors, Chancay culture, Central Coast, A.D. 1200–1400. Yale University Art Gallery, New Haven)

Fig. 3.19 Reconstructed thatched roofs on a masonry structures at Machu Picchu. (Author's photograph)

Fig. 3.20 Contemporary street with a central water channel in Cusco. (Author's photograph)

Betanzos (1996 [1576], pp. 50–58, 69–73) discussed the organization Pachacuti brought to the Cusco valley and Zuidema (1990a) described a system of *chapas* used to divide responsibility for the lands around the city. Each of 10 Inca lords was granted one of the 10 chapas of land surrounding Cusco through the mountainsides. Produce from the villages on the land would then be collected in storehouses erected for that purpose. Chapas were a fourth method used by the Incas for society organization, the others being panacas, ayllus, and the calendar. Chapas were distinguished from panacas and ayllus as they had separate functions (Zuidema 1990b).

Taxpayers in the empire were organized by a decimal system. A *Pachaca* was a group of 100 taxpayers and a *Huaranca* was a group of 1000 that included 10 Pachaca. A *Huno* was comprised of 10 Huaranca and represented 10,000. Pachacas were divided into two groups of 50 and further into groups of 10. A *huamani,* or province, contained 40,000 families. Officials were appointed at each level of administration (Hemming 1970 ; Zuidema 1964 ; Zuidema 1990a).

3.10 Succession

Succession as ruling Inca normally passed to the eldest primary son. However, if this son was thought to be weak or incompetent, he might be challenged and deposed by a stronger brother. Royal blood was important to the Incas, but they also valued powerful leadership and accepted the occurrence of these struggles for control. Several of the eleven historical Incas are thought by some to have acceded to the throne only after such an event (Hemming 1970).

Because the son did not inherit his father's property or riches when succeeding him, this created a need for continued conquest as each new Inca sought to establish his fortune. A significant portion of the land in Cusco was occupied by palaces and estates of the Incas, or their mummies, and royal family members of the panacas.

3.11 Festivals

Pachacuti instituted a ritual calendar that was measured by the position of the Sun on the horizon. Religious and agricultural ceremonies and festivals were celebrated and served to reinforce the legitimacy of the emperor and the ruling elite. Rituals marked times for planting and harvesting, as well as major religious celebrations of the Sun. Examples of prominent festivals were those of Inti Raymi at the June solstice, Capac Raymi Quilla at the December solstice, Pacha Pucuy Quilla at the March equinox, and Coya Raymi Quilla at the September equinox.

Inti Raymi was the Incas' winter solar festival at the time of the June solstice and it is still celebrated in Cusco today. Garcilaso (1961 [1609], p. 217) wrote that the feast of Inti Raymi was the most important one of the year. It was a festival for the masses that brought many pilgrims to Cusco (Dearborn, Schreiber, and White 1987). It was an elaborate ritual that took place over many days where thanks were given to the Sun and prayers said for the crops. Inca citizens chanted throughout day, the volume of their voices rising as the Sun rose higher and falling when it descended again (Zuidema 1986).

Capac Raymi was a summer festival of the Sun at the time of the December solstice. This was a celebration of crop germination and the start of a new season. It also was important as the annual time for young Inca adolescents to undergo the rituals of coming-of-age (Hemming 1970). Capac Raymi was principally celebrated by the nobility rather than the masses (Dearborn et al. 1987).

3.12 Climate

The cold Humboldt Current flows northerly along South America's Pacific coast and is responsible for upwelling an abundance of marine life off the shores of Peru. The cooling influence of the current extends to the surrounding marine air, hampering precipitation and creating a very arid coastal climate.

Conversely the climate further inland and higher in the Andes experiences great amounts of rainfall during annual rainy seasons. An additional biome existed for the Incas in the low Amazonian rain forest below the Andes' eastern slopes.

The annual rainy season in Cusco extends basically from October to April, followed by a dry season from May to September. The Incas coordinated the planting of maize with the beginning of the rainy season and harvested their crop as the dry season approached (Urton 1981a). The comings of such agricultural events were tracked by following the position of the Sun on the horizon (Fig. 3.21).

Weather patterns were occasionally interrupted by what we know as El Niño. Andeans learned to predict its coming through observance of the relative brilliance of the Pleiades asterism after its heliacal rise in early June. Orlove, Chiang, and Cane (2000) have studied this method used by the Incas and say that it may have existed for four centuries.

The Incas maximized efficiency of their agricultural environment through cultivating produce optimal for specific environmental zones. Certain crops thrived at high altitudes, while others needed to be grown much lower. Cooperative efforts exchanged foods between these regions. Astronomical agricultural calendars varied for proper planting and harvest in each of the zones (Urton 1981a).

Fig. 3.21 Two large solar pillars above Q'espiwanka in Urubamba can be seen as small points to the left of the Sun's center on the Cerro Saywa Ridge and were used to mark the Sun's position at and near June solstice; similar pillars that surrounded Cusco were destroyed in the Catholic extirpation of idolatries. (Author's photograph)

3.13 Agriculture

The importance of the Sun for agriculture in the Andes was known long before the Incas. The Incas pursued agriculture aggressively and achieved considerable success. Agricultural practices freed the general populace from the need for subsistence farming and thus many of them became available for military service and construction projects (Bauer and Dearborn 1995; Niles 1987).

Agricultural terraces greatly benefitted the Inca Empire and they used them extensively (Fig. 3.22).

Inca engineers adopted the concept from the Huari, who in turn had discovered the technique from earlier cultures (Wright and Valencia 2000). These terraces were designed to maximize crop productivity and efficiency in a mountainous environment. Not only were they quite effective in this regard, but terraces also served to protect against erosion and assist in irrigation. Mountains were terraced throughout many parts of the empire, enabling cultivation in fertile topsoil in areas that would otherwise have been untillable (Niles 1987).

Planting required many decisions as to its timing. Observations of both the Sun and Moon were used to determine optimal dates for sowing. Maize was a major nutrition source and was the most significant crop in the empire. It is thought that pillars were built on Cusco's horizon to establish the proper time for planting (Bauer

Fig. 3.22 Terraces below Pisac. (Author's photograph)

Fig. 3.23 Chicha is still brewed today, both with the traditional maize and in a newer version made with strawberries. (Author's photograph)

and Dearborn 1995; Urton 1981a). Maize was also the primary ingredient in *chicha*, the corn beer widely consumed throughout the empire that also figured prominently in religious and state ceremonies (Fig. 3.23).

The growing season for maize is from October to May, with the fields first plowed in August. Cobo (1990 [1653], p. 144) described a ritual planting in August to initiate the maize season that took place in a *chaucara*, or field, east of Cuzco called Sausero.

Large tracts of land were set aside to till maize in support of members of cults of Inca deities such as the Sun, Moon, stars, thunder, and lightning. Capac Raymi and Inti Raymi, the two of the most significant Inca festivals, celebrated the planting and harvesting of maize near the respective times of the December and June solstices (Bauer and Stanish 2001).

Potatoes were another staple of the Andes and ultimately spread from the region throughout the world.

Coca was revered for its stimulant qualities. Its leaves were chewed to produce an effect that minimized hunger and fatigue. It was grown primarily on the eastern side of the Andes far below the Incas' primary habitat and found use in many religious ceremonies (Hemming 1970).

3.14 Irrigation

The Incas were masters of hydraulic engineering, expertly irrigating their agricultural terraces and fields. There are many extant examples of fountains and canals at numerous archaeological sites in Peru (Fig. 3.24).

Fig. 3.24 Canal at Ollantaytambo. (Author's photograph)

The Incas designed fountains to be both practical and ceremonial and they likely served as well to facilitate the life-force energizing effects of camay. The series of 16 fountains at Macchu Picchu appear to be quite practical with regard to consumption, while fountains at Ollantaytambo and Tipon seem more ceremonial in nature (Figs. 3.25 and 3.26).

Canals were fed by streams, springs, and reservoirs. Some were large while others quite narrow and focused with means of diverting water flow to selected branches. Many were lined with stone or carved into walls, often between 40 cm and 80 cm in width. Terraces were irrigated through such channel systems and sometimes included parallel channels to back up the first should it become nonfunctional (Niles 1987).

3.15 Imperial Expansion

Starting with Pachacuti, the Incas began a great period of conquest and expansion continued by his son, Topa Inca, and grandson, Huayna Capac. In a matter of decades, the Incas subjugated peoples and territory spanning over 4800 km along the Andes, all without benefit of the wheel or horses.

The Incas preferred to assimilate tribes willingly rather than by force. Many complied while others were ultimately compelled to submit. Certain practices of

Fig. 3.25 A section of the fountains at Machu Picchu. (Author's photograph)

these conquered peoples were adopted by the Incas, an example of which is exhibited by the similarity of masonry techniques between those of the Incas and what has been found of the Tiwanaku near Lake Titicaca.

The Incas were adept at dealing with newly assimilated subjects and were tolerant of their local religious practices, providing that they first acknowledged the superiority of the state solar worship. Recently acquired idols, along with priests to attend them, were transported to Cusco and honored with proper respect, while at the same time providing collateral for the willing cooperation of their associated tribes (Bauer and Stanish 2001; Hemming 1970).

3.16 Pilgrimage

Pigrimage existed in the central Andes long before the rise of the Incas to power. Early examples in the time of the Huari and the Tiahuanaco can be found in the temple of Pachacamac (Fig. 3.3) on the coast near Lima and on the Isla del Sol in

Fig. 3.26 Each fountain at Machu Picchu successively descends a steep slope. (Author's photograph)

Lake Titicaca (Fig. 2.11). These sites figured prominently in Andean mythology and cosmology and became focal points for ritual movements (Zuidema 2008a).

Certain pilgrimages to state shrines were instituted to reinforce values of Inca society and religion and to serve as a form of indoctrination for the legitimacy of the ruling elite. The Incas found it imperative to ideologically assimilate tribes after conquering them militarily (Bauer and Stanish 2001). A pilgrimage center often contained what was sacred to the pilgrims who traveled there to worship (Silverman 1994).

The Incas became experts at influencing the thinking of their citizens. As pilgrims neared the shrine, their experience each step of the way was carefully orchestrated by state and religious officials. Tenets of the solar cult were instilled and association of the Inca as the son of the Sun was paramount. Each aspect served to carefully exert control over the perceptions of the population. The experience allowed persons making the pilgrimage to transform from residents of mountain villages to members of the greater empire (Bauer and Stanish 2001).

Zuidema (2008a, b) discussed three other forms of pilgrimage that were practiced along the southeast to northwest axis of the solstices that extended between Vilcanota to the southeast and Ollantaytambo to the northwest. He called these three ritual movements (1) a procession, (2) a pilgrimage, and (3) a race (Fig. 3.27). The first two were conducted by Tarpuntay priests near the time of the June solstice

Fig. 3.27 Routes of travel for the procession, pilgrimage, and race. (From Zuidema 2008b, p. 257)

while the third involved younger men, perhaps of more common status (Zuidema 2008a, b).

The procession took place daily during this period. Priests traveled from Huanacauri on the southeast side of Cusco to Quiancalla on the northwest. They sacrificed lambs at their journey's beginning at dawn, its end at sunset, and along the way at the Coricancha at noon (Zuidema 2008a, b).

In the pilgrimage, another group of Tarpuntay priests traveled southeast from Huanacauri and Mutu to Vilcanota. Their round-trip journey lasted for a month as they worshipped the Sun while moving through the mountains to Vilcanota, with return by the Vilcanota River and Quispicancha, or Tipon. The process was to help the Sun turn around at the June solstice (Zuidema 2008a, b).

Zuidema said that a foot race called *Mayucati* complemented the pilgrimage. In the race, young men competed after the December solstice by running alongside sacrificial ashes cast upon the Vilcanota River as they flowed northwest to Ollantaytambo. They then raced each other through the mountains back to

Fig. 3.28 A double-jambed doorway at Machu Picchu. (Author's photograph)

Quiancalla and Cusco in an event set to end at the time of the February zenith passage (Zuidema 2008a, b; Fig. 3.28).

3.17 Building an Empire

During the period between Pachacuti's conquest of the Chanca and the beginning of the Spanish invasion, building and construction proceeded at a rapid pace. Workers freed by surpluses in agriculture were used for both conquest and construction while artisans and their building techniques were assimilated from throughout the realm (Hemming and Ranney 1982).

An essential element of this massive building campaign was that of the use of mitmae labor from across the empire. This system provided temporary labor on a rotating basis and worked well in a society without money. Major public works projects such as buildings, palaces, agricultural terraces, canals, bridges, and roads were all constructed through the use of such work forces. Village men were required to take part in the rotations as a service to the state and the Incas were careful to relocate them to work in compatible climates and elevations. Once their period of service was over they returned to their regular homes and were replaced by new mitmaes from another area (Gasparini and Margolies 1980; Niles 1987).

3.18 Architecture

Every new emperor acceding to the throne built his own palace and monuments. All Incas had palaces in Cusco maintained by their panacas long after their deaths. Pachacuti, Topa Inca, and Huayna Capac were no exceptions as they constructed fine palaces in the capital and other estates in the countryside. Niles (1999) argues that these architectural examples were manifestations that require knowledge of Inca history to truly appreciate.

Royal architecture was a means used by Inca emperors to enhance their status as rulers and visibly represent their inclusion in the dynastic succession of kings since Manco Capac. Rules of succession left a new emperor none of his father's possessions and required him to build his own palaces and country estates and amass his own wealth, a system that allowed for great expression by each new Inca while establishing his relative stature.

Most civil works of the Incas were built by Pachacuti, Topa Inca, and Huayna Capac during a 77-year period between 1450 and 1527. Pachacuti established the empire's architectural style and spread it throughout his realm (Gasparini and Margolies 1980; Christie 2007). Topa Inca followed his father's basic model, but with innovations of his own. His son, Huayna Capac, did likewise.

The Incas did not normally rebuild atop earlier ruins, instead choosing original sites for their construction. The degree of craftsmanship in the masonry used in a building directly indicated the structure's relative importance. The finer the cut and fit of the stones, the higher the stature of the building. Double-jambed doors denoted entries for use only by elites. Construction variances sometimes existed due to building location or function (Hemming and Ranney 1982; Fig. 3.28).

Inca masonry was generally of two types, polygonal and coursed, and both are quite precise without benefit of mortar (Paternosto 1989). In polygonal masonry, random interlocking faces of stones were cut and polished to fit together with precision, but no two walls were the same. Coursed masonry consisted of polished surfaces laid in precisely fitting and orderly horizontal rows, each successive row slightly smaller than the one below. Examples of stone types used include andesite, diorite, porphyry, granite, and limestone. The Incas were fond of trapezoidal doorways and niches instead of using arches. These trapezoidal constructs were a hallmark of the Incas and their empire (Gasparini and Margolies 1980; Hemming and Ranney 1982; Fig. 3.29).

Niles describes that building proportions were standard in Inca designs. This is in keeping with other Inca precision, and she claims a ratio of short to long sides between 1:1.71 and 1:2.07. Structural size was important for status and prestige and

Fig. 3.29 An example at Machu Picchu of a coursed masonry wall carved to fit with great precision without mortar. (Author's photograph)

significant buildings often incorporated open spaces to increase the effect (Niles 1987).

Inca structures were almost always single-story with thatched roofs. When a second story did exist, the stairway was often outside. Common residences were primarily single-room and buildings with multiple rooms generally did not connect them internally. Windows were less common and the majority of extant second stories were added by the Spanish after the conquest (Gasparini and Margolies 1980).

Inca buildings normally followed a standard rectangular plan established during Pachacuti's reign that was employed in distant conquered territories as well as in Cusco. Doors were generally located on one of the rectangle's long walls, but side doors were sometimes added to the largest of buildings (Gasparini and Margolies 1980).

The Incas also used a circular plan for certain buildings not used as residences. Storehouses were often circular, as were funerary structures called *chullpas*. Examples of curved walls not completing a circle can also be found at sites such as the River Intihuatana (Fig. 3.30), Machu Picchu, Cusco, Tipon, and Pisac (Gasparini and Margolies 1980). The Sunturhausi in Cusco was circular with windows, a high roof, and a mast. Its astronomical uses are described in Sects. 6.6 and 6.8. Figure 3.31 is a drawing of the Suntuhuasi by Guaman Poma.

Fig. 3.30 The open curved wall of the "tower" at the site of the River Intihuatana near the river below Machu Picchu. (Author's photograph)

Canchas were walled rectangles surrounding a number of one room structures designed for a similar purpose. Kallankas were great halls with many doorways intended for festival observance in inclement weather (Niles 1999).

Niches are most commonly found on the inside of building walls. Niles (1987, p. 215; Fig. 3.32) found them to be symmetrical on walls facing doorways and many of a standard size close to 80–90 cm in height and 1.25 m above floor level, however some were much larger to accommodate mummies. Others were symmetrical on opposing walls. As with trapezoidal doorways, stone lintels were placed to complete the tops of niches and those found on exterior walls often exhibit double or triple jams as signs of status. Niches were frequently used to display items of religious veneration.

3.19 Inca Roads

The Incas built an impressive system of roads for movement throughout their empire. These thoroughfares were constructed of carefully fitted stones and spread throughout the realm in a network extending for a total of over 16,000 km. Most

Fig. 3.31 Drawing depicting the Sunturhuasi. (Guaman Poma, f. 329)

roads were built higher on the sides of mountains, thus placing them in the best positions for travel during the rainy season and keeping them apart from land that was prime for agriculture (Figs. 3.33 and 3.34). The network was critical for rapid travel of the emperor's armies and also in maintaining control over conquered territories. Administrative centers were instituted systematically along the many routes established, particularly to the north toward Quito (Hemming and Ranney 1982; Niles 1987).

The most important highways were reserved for official use and the main roads leading to the four suyus of the empire all radiated from Cusco. The network extended north to Quito, west to the Pacific, and south past Lake Titicaca. These roads were sometimes used as well to establish and designate boundaries. The system was constructed and maintained largely by mitmae labor (Gasparini and Margolies 1980; Hemming and Ranney 1982; Niles 1987).

Roads were meant for foot travel as the Incas had no horses or wheeled vehicles. Llamas could be used as pack animals but were not suitable for riding.

Fig. 3.32 Niches in a wall at Tipon. (Author's photograph)

Fig. 3.33 Example of a section of Inca trail near Machu Picchu. (Author's photograph)

Fig. 3.34 The Inca trail here is the green horizontal line (vegetation) across the cliff face; this leads to the left to what is known as "the drawbridge" (Fig. 9.3) and an alternate entrance to Machu Picchu. (Author's photograph)

As carts were impractical on these mountain highways the Incas never had the need for the wheel. Llamas, however, were well suited for transporting goods on their backs along the often vertical Inca trails. The many bridges required throughout the empire were normally engineered as spans by rope suspension (Hemming 1970).

3.20 Carved Rocks

Examples of carved rock sculptures can be found throughout much of the Inca Empire. Emperors felt it their right to improve upon nature by sculpting in situ outcrops that often became huacas. Improvements to rocks appear to have been state-controlled and likely guided by a certain class of artisans as evidence does not suggest innovation. These methods also were not sudden inventions, but instead had developed over time with knowledge acquired from other societies (Niles 1987; Paternosto 1989).

Pachacuti commanded a study of techniques of the Tiwanaku, a likely source of Inca interest and expertise in masonry. The Incas had only soft metals but found that they were able to shape blocks with harder stones to achieve desired results. Sculpture was a religious art form as well as practical and new examples were quite prevalent during Pachacuti's reign as emperor (Gasparini and Margolies 1980; Paternosto 1989; Van de Guchte 1990).

Paternosto (1989) tells us that geometric symbols found carved in stone often originated first in textile designs. Primary cosmological figures such as pumas, serpents, and condors began that way, as did the human form.

Limestone was commonly selected for carving, but Van de Guchte (1990) also describes examples where diorite was used at Suchuna, andesite at Rumihuation and Saihuite, black granite at Choquequilla, and porphyry at Ollantaytambo. He also mentions a division of labor with specialized workers trained for such as stone cutting, carving, and construction.

Van de Guchte has identified three styles of rock carvings: severe, composite, and figurative. The severe is characterized by geometric cuts, such as those found at Kenko, Sacsahuaman, Chinchero, and Machu Picchu. Composite style includes carvings in high relief as evidenced at Lacco and Chinchero. Figurative style carvings depict plants and animals in both high and low relief, with Saihuite holding a fine example.

The Incas used carved rocks as a vehicle for promoting state ideology and the solar religion. They were symbols of commemoration, mediation with the cosmos, and state identity, all the while remaining part of the Incas' perception of their sacred relationship with nature and the land (Van de Guchte 1990).

A rock, once carved, became a hierophany and was worshipped by the Incas in a way unfamiliar to many not of the Andes. Embedded in the Earth, these sculpted manifestations of the sacred were connected with the powers of the underworld and became venerated when enhanced by elaborate carving. Sculpted outcroppings, as huacas, were an important part of cosmological symbolism regarding the three worlds of the Incas (Paternosto 1989).

Examples of reverence for stone are prominent throughout Inca culture. In Inca mythology the first beings emerged from places such as caves, rocks, and springs, all connected with the underworld. Each king had an effigy called a *huauque*. These "stone brothers" were treated similarly to their likenesses and "possessed" houses, fields, and servants. The huauque was a companion to the emperor and provided advice as required (Paternosto 1989; Van de Guchte 1990).

Ritual stairs are a dominant motif, perhaps expressing movement from the underworld to the Earth to the heavens and are often associated with these three realms. Carved stairs are common and frequently non-functional, such as those on inaccessible cliff sides at Ollantaytambo and within a cave at Machu Picchu (Fig. 3.35). They normally include three steps, corresponding with the three worlds, and likely were symbolic representations of this cosmology. Carvings of condors, pumas, and serpents proliferate as representatives of these respective spiritual domains.

The *chakana*, or Andean square cross of three levels, is another stepped symbol representing transition between the three worlds. Chakana, in Quechua, means bridge, or to cross from one place to another. Sculpted on each of the four sides of

Fig. 3.35 Ritual stairs at the base of The Torreon at Machu Picchu. (Author's photograph)

the cross are surfaces resembling three stairs, again the common motif of the underworld, our world, and the heavens (Figs. 3.36 and 8.23).

Seat carvings are common and Van de Guchte (1990) believes they may have been meant to be altars as well as chairs (Fig. 3.37). I found many to face in the directions of significant sunrises or sunsets.

3.21 Summary

To gain a true appreciation of Inca astronomy, we must first gain an understanding of Inca culture and the way its people viewed the world. We have to be careful not to make the same mistake as the sixteenth century Spaniards by interpreting what we see through our own frame of reference.

The Incas had a special relationship with the Sun. It formed the basis of their religion and they believed their royal ruler to be the son of the Sun. Their veneration in this regard was something for which we have no equivalent. The Moon also was viewed with great reverence and was thought to be both the wife and sister of the Sun and, by extension, the mother of the queen. Daily movements of these bodies were observed and tracked closely.

Solar worship helped to reinforce the divine authority of the emperor, the son of the Sun. Shrines and pilgrimages were created to ensure the population understood without question the royal relationship between the Sun and the ruling elite.

Fig. 3.36 The general form of the chakana. (Huhsunqu / CC BY-SA / https://creativecommons.org/licenses/by-sa/2.5)

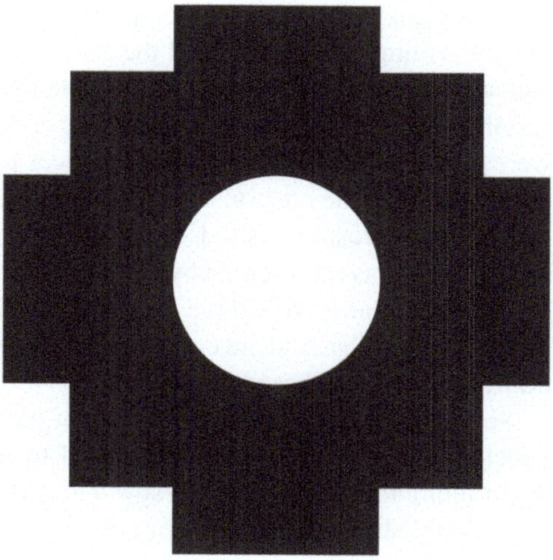

Fig. 3.37 Possible seat carvings at Chinchero in the Sacred Valley. (Author's photograph)

Mountains and rock were worshipped for the spiritual powers held within them. Deities within snow peaks held the power of the Earth's life-giving force of water and supplied it in the cosmological cycle to the people living below them. Camay was thought to bring life and animation to otherwise inanimate objects. Outcroppings were carved at the direction of ruling Incas to further energize them while enhancing the work of the creator.

The Incas' cosmos included three worlds, the below, the here and now, and the above. Each was represented by an animal, respectively the serpent, puma, and condor. Carved stairs with three steps representing the three cosmological Andean worlds and similar Andean crosses are frequently occurring motifs symbolizing the transition between these worlds.

Solar regulation of agriculture was of paramount importance. Pillars were constructed and other means were devised to make use of the Sun's position when determining the proper times to sow and harvest specific crops. Festivals of the Sun were established to commemorate these dates, which also served to reinforce the belief that the Inca had a special relationship with the solar deity.

Many huacas were carved from outcroppings and other stones and survive to this day. The Incas believed that carving and camay energized a spiritual force giving huacas animation and life. I found many astronomical orientations to exist within the sculpting of such stones and will discuss them in later chapters. The Incas did not view astronomy as a separate entity, but instead as an integral part across many components of their culture. It is only through an understanding of such complex interrelationships that we can begin to comprehend fully what the Sun, Moon, planets, and stars meant to the peoples of the Andes.

Chapter 4
Ceques and Huacas

Fig. 4.1 Carved rock huaca at Chinchero. (Watercolor by Jessica Gullberg)

Huacas and many of their functions were previously introduced in several sections of Chap. 3 (Sects. 3.1, 3.2, 3.3, 3.4, 3.7, 3.8, and 3.20). This chapter describes the Cusco ceque system and the role that huacas played in in it to organize society.

Tom Zuidema, of the University of Illinois at Urbana-Champaign, wrote *The Ceque System of Cuzco: The Social organization of the Capital of the Inca* (1964). He was a preeminent scholar of Inca astronomy and society and has been honored by the Peruvians with an exhibit dedicated to him that is located within the

© Springer Nature Switzerland AG 2020
S. R. Gullberg, *Astronomy of the Inca Empire*, Historical & Cultural Astronomy,
https://doi.org/10.1007/978-3-030-48366-1_4

Coricancha. Zuidema tells us that the shrines of Cusco were part of a great system devised to organize Inca society and religion, as well as give order to astronomy and the calendar. The ceque system was a "device for integrating astronomy, cosmology, and sociopolitical structure" (Zuidema 1981c, p. 169). In the article "Catachillay: The Role of the Pleiades and of the Southern Cross and α and β Centauri in the Calendar of the Incas," Zuidema (1982a) described a two-dimensional, walk-through calendar of 41 *ceques*, or lines, radiating from Cusco that collectively contained 328 *huacas*, or shrines, some of them carved outcrops. A number of those extant have been found to have astronomical orientations.

4.1 History

The Incas venerated many features of the natural landscape thought to be endowed with meaning and sacred power. These became *huacas* and prior to the Spanish invasion there were many hundreds of smaller ones. Bray (2015) characterized this nicely as "…the wholeness of wak'as seems to have extended beyond their corporality or materiality to encompass the broader field of relations within which they are embedded – an aspect that may, in fact, have figured into their 'holiness'" (p. 5). Major huacas required maintenance and caretaking. Gifts were made to the powers of the shrines. Animals and produce were sometimes sacrificed to the huaca and used to support the attendants. The huacas surrounding Cusco were organized in lines or *ceques*. Pachacuti was concerned about the lands held by mummies and thus redistributed land and water rights in a manner organized by this ceque system (Betanzos 1996 [1576], pp. 69–70; Sarmiento 2009 [1572], pp. 132; Fig. 4.2).

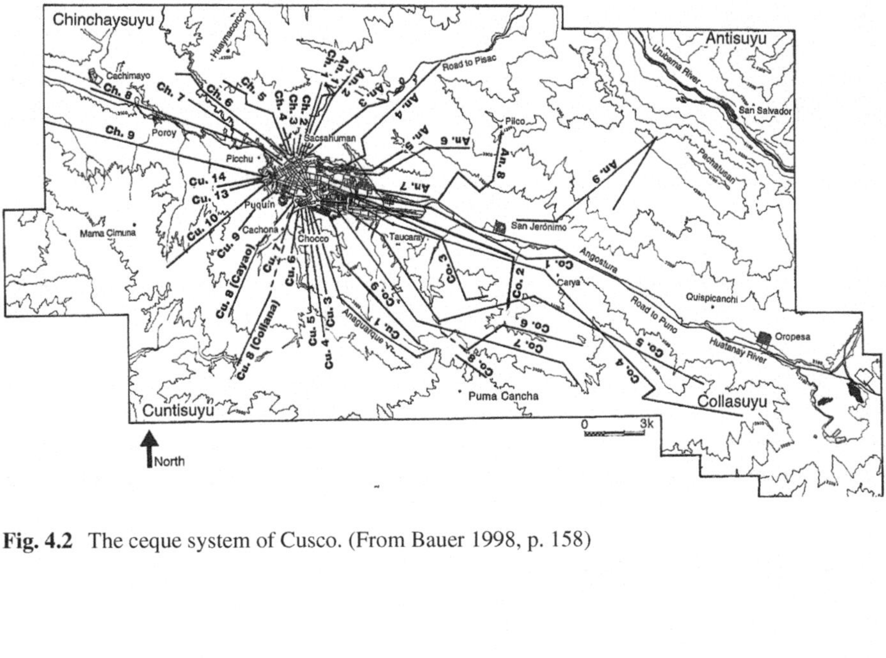

Fig. 4.2 The ceque system of Cusco. (From Bauer 1998, p. 158)

The Jesuit priest, Bernabé Cobo, published a list of huacas in the vicinity of Cusco. In *Historia del Nuevo Mundo* he gave the names, descriptions, locations, and offerings of each. Cobo (1990 [1653], pp. 51–84) provides a list of 328 huacas. Of that total 261 are natural features of the landscape, 89 involved water, 82 were geographic features, 83 were rocks, and 2 were botanical (Niles 1987).

According to Cobo, the huacas were organized along ceques, partially to regulate their maintenance and to organize sacrifices at proper times (Rowe 1980). The shrines were arranged on ceques and registered on *quipus* (Sect. 3.8), which ordered them in a manner consistent with the highly organized administration of the Inca kingdom. The huacas were arranged outward from the Coricancha along four groups corresponding to the four administrative quarters of Tawantinsuyu. Ceque lines did not overlap so that progression from one shrine to the next along a particular ceque was a straightforward matter. There were 42 ceques, but the last two were grouped together as one leaving 41 ceques in all (Zuidema, 1983; Fig. 4.3). The first ceque

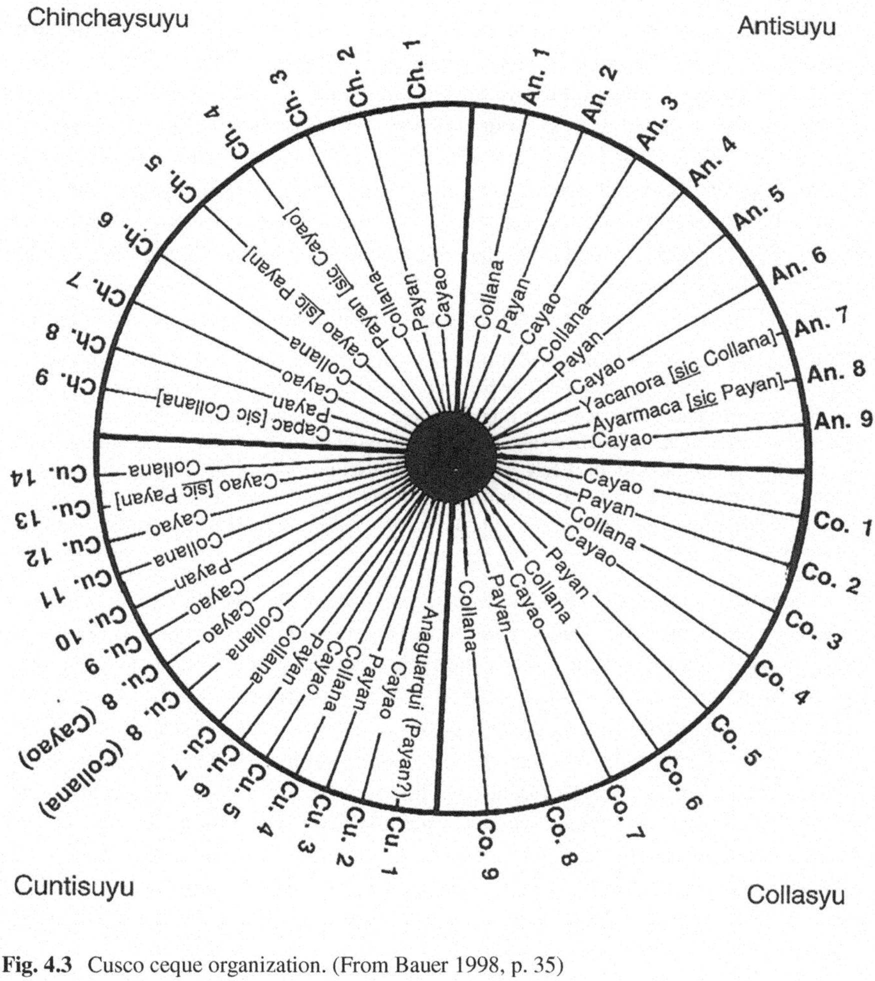

Fig. 4.3 Cusco ceque organization. (From Bauer 1998, p. 35)

was royal and called capac, thus there were 40 non-royal ceques and one royal one. The huacas were not placed in a geometrically regular sequence along the ceques, they were not equidistant. They were typically within a half-day walk from the Coricancha. The ends of many ceques were reported to be where one would lose sight of Cuzco (Niles, 1987).

Cobo was not the original author accounting for ceques and huacas, however. An earlier manuscript called *Relación de las Huacas,* whose author is not certain, was written c. 1569. Cobo appears to have relied extensively upon this text while compiling his own. Zuidema (2008b) felt that this original author was Juan Polo de Ondegardo.

4.2 Ceques

Ceques were important features of the shrines in the Cusco valley, and one of their symbolic roles may have been to affirm and supplement the inherent directionality of huacas. In a way they can be reminiscent of the famous Nazca lines, for which there are a variety of interpretations such as water rituals, astronomical sightlines, and depictions of star patterns. Ceques were organizational in intent, indicating sequences of ritual visits, responsibilities for individual panacas and ayllus, and assignment of territory and irrigation sources. Sherbondy (1992) says that the Incas encoded the responsibilities for huacas "into a map whose physical representation was probably a quipu" (p. 60). She also states that the ceque system codified water rights for panacas and ayllus.

Cobo (1990 [1653]) speaks of the ceques prior to his listing of the 328 huacas:

> From the Temple of the Sun, as from the center, there went out certain lines which the Indians call *ceques;* they formed four parts corresponding to the four royal roads which went out from Cuzco. On each one of those *ceques* were arranged in order the *guacas* and shrines which were there in Cuzco and its region like stations of holy places, the veneration of which was common to all. Each *ceque* was the responsibility of the kinship units and families of the city of Cuzco, from within which came the attendants and servants who cared for the *guacas* of their *ceque* and saw to offering the established sacrifices at the proper times. (p. 51)

The huacas surrounding Cusco were arranged on these lines. Ceques helped to divide the region into the four suyus emanating from Cusco (Fig. 4.3). Zuidema (1964) found the ceque system to be a significant component of social classification. He also maintained that it was useful for calendrical purposes and the accomplishment of tasks at specific times throughout the year (Zuidema 1981b).

When viewed on a map from above, the ceque system resembles a quipu (Sect. 3.8) radiating from Cusco's Coricancha. This may be more than just coincidence. The knots of quipus were used as mnemonic aids and huacas on the ceques served also to facilitate remembrance (Niles 1999; Paternosto 1989; Fig. 4.4).

Cusco society was divided into two moieties and each moiety into two suyus. Ceques appear to have delineated these geographical divisions and the modern

Fig. 4.4 An example of a quipu. (Claus Ableiter nur hochgeladen aus enWiki / CC BY-SA / http://creativecommons.org/licenses/by-sa/3.0/)

cataloguing of ceques is organized by the suyus in which they lie. Hanan Cusco encompassed Chinchaysuyu to the northwest and Antisuyu in the northeast. Hurin Cusco included Collasuyu to the southeast and Cuntisuyu in the southwest (Zuidema 1964).

Zuidema (1964) argues that ceques marked the route of sequential offerings to adjacent huacas. Some of the ceques may also have served as pathways and convenient routes of movement. Instead of simple trails for ritual-pilgrimage, ceques may have established administrative socio-political boundaries and may have at times followed existing roads. Both interpretations of ceques, as social and administrative delineators and as symbolic geometrical and astronomical alignments, seem to have merit. Some ceques marked sightlines with astronomical intent, while others were used to organize and indicate the responsibilities for the care and maintenance of huacas provided by individual panacas and ayllus.

Zuidema (1964) describes ceques in four groups that correspond with Cusco's four administrative quarters. Nine ceques fall within Chinchaysuyu, nine within Antisuyu, nine within Collasuyu, and the remaining fourteen in Cuntisuyu. Within the suyus they were ordered primarily in groups of three and the names *Collana, Payan, and Cayao* were assigned, one to each in a triad of ceques. The ceques of Chinchaysuyu were ordered counterclockwise while all the others were clockwise in progression (Zuidema 1964; Fig. 4.3).

Zuidema (1964) then describes further the social class association with these ceque delineations. Collana, Payan, and Cayao refer to panaca and ayllu status and Zuidema (1983) states that Collana was the most prestigious and was made up of members of pure Inca descent. Payan included subsidiary kin such as offspring born from a Collana man with a non-Collana woman. Cayao comprised the remainder of the population that were not of Inca descent. No more than one panaca was assigned to each ceque cluster (Bauer 1998; Zuidema 1964).

The ceques within a suyu were sequentially numbered and Bauer (1998) refers to them by this number along with the suyu's abbreviation (Fig. 4.3). The first ceque of Chinchaysuyu is therefore known as *Ch. 1* and the last one *Ch. 9.* Bauer (1998) says that the ceque Ch. 1 "[defined] the division between Chinchaysuyu and Antisuyu" (p. 49). He continues that Ch. 9 "marked the division between Chinchaysuyu and Cuntisuyu, and the western boundary between Hanan and Hurin Cusco" (p. 49). Zuidema (2002), however, argued that Bauer was incorrect with regard to ceque borders "as then those ceques would have coincided with bordering ceques of the next suyus" (p. 593).

Polo de Ondegardo (1965 [1571]) and later Cobo (1990 [1653]) related that every village had ceques. While the ceque system of Cusco is the only one to have been extensively substantiated, the preponderance of huacas surrounding Machu Picchu with certain alignments suggests there may have been a ceque organization employed in this area as well. One such alignment connects the Sacred Plaza of Machu Picchu with the Sun Temple of Llactapata along the axis of June solstice sunrise and December solstice sunset (Fig. 9.2).

Salazar (2004) notes the large number of shrines at Machu Picchu and relates them to the huacas that surround Cusco. She suggests that Pachacuti invested a large amount of skilled labor in establishing these because of the special association he felt with the "supernatural forces immanent in the landscape and the celestial sphere," and that his connection with these forces needed to be "actively reaffirmed through daily ritual" (p. 41; Fig. 4.5).

Below Machu Picchu, not far from the confluence of the Aobamba and Urubamba rivers, lies a large and complex shrine initially identified by Bingham as the Intihuatana of the Urubamba River, which may be part of a ceque (Fig. 9.33). Its massive carved granite stone is approximately in line with the June solstice sunrise as viewed from the Llactapata Sun Temple and is discussed in Sect. 9.3.

4.3 Huacas

Many features of the Andean landscape were venerated by the Incas as they felt them to be endowed with supernatural powers. Cobo (1990 [1653]) stated that the Incas venerated large trees, roots, springs, rivers, lakes, hills, and mountains. He continued "They also did reverence to these places and made offerings" (p. 44), and that they worshipped anything natural that was perceptibly different. "All of these idols were worshipped for their own sake, and these simple people never thought to search or use their imaginations in order to find what such idols represented" (p. 45). Huacas were systematically worshipped and cared for and were integral parts of Inca religion and culture. They often were shrines to ancestors who, it was believed,

Fig. 4.5 Machu Picchu. (Author's photograph)

could influence the living. The most powerful huacas required maintenance, caretaking, and offerings.

Salomon and Urioste (1991) state that "a huaca was any material thing that manifested the superhuman: a mountain peak, a spring, a union of streams, a rock outcrop, an ancient ruin, a twinned cob of maize, a tree split by lightning" (p. 17). Mummies could become huaca, as well as could the children sacrificed in *Capac Hucha* by being buried alive.

Huacas were part of the officially organized worship of the Inca capital, and Cobo's account provides insight into what things were considered sacred. Sacrifices and offerings were made at the shrines with such as gold, silver, clothing, sea shells, and sheep. Extraordinary offerings involved the sacrifice of children, human and animal figurines, coca, llama blood, and firewood dressed as people (Cobo 1990 [1653], pp. 109–114). Winding or zigzagged channels were carved into certain huacas, perhaps for the flow of chicha or blood which may have been used for divination (Figs. 3.9, 7.9, and 8.8). Following the conquest, the Spaniards began a campaign against indigenous religion that included the extirpation of huacas (Sect. 2.6). Some huacas, mainly large carved rocks, could not be eradicated and remain extant.

Two shrines marked the spot where travelers would first lose sight of Cusco and where they were given coca for a safe journey. Direct views from huacas were important, not only for the first and last views of Cusco by travelers, but also for the replication of form (Niles, 1987). According to Cobo, several huacas were significant because they resembled the form of a sacred mountain, the human form, or a creature such as a serpent, puma, and condor (Fig. 4.6).

Huacas often played a role in Inca festivals with *huacacamayocs*, or huaca specialists, coordinating shrine worship and offerings. Huacas were holy places and varied from major state shrines to those worshipped by a single household (Bauer 1998; Bauer and Stanish 2001; Niles 1987).

Certain huacas were thought to be oracles with powers to counsel the emperor. The huacacamayocs acted as facilitators for messages between the oracle and their king. Huacas were focal points of communication with the Pachamama by virtue of their physical connections with the Earth (Hemming and Ranney 1982; Van de Guchte 1990).

Cobo (1990 [1653], pp. 51–84) provided a comprehensive list and description of all 328 huacas of the Cusco ceque system. He also mentions that the Coricancha was counted as an extra huaca.

Brian Bauer wrote *The Sacred Landscape of the Inca: The Cusco Ceque System* (1998), and used Cobo's descriptions when he developed his huaca classification system employing such designations as Ch 1:2 for Kenko Grande. This is the classification method that is followed in this book.

Fig. 4.6 A replication at Machu Picchu. (Author's photograph)

Carved rock huacas are bedrock features with their roots in the Earth, an important aspect of symbolism involving the three Inca worlds. The Incas viewed them as stone deities (Bray 2009). Pachacuti believed he could modify and enhance the work of the creator, thus the carving of rock huacas may be attempts to express the inherent meaning contained in a unique rock, its "endogenous" meaning (Paternosto 1989). Replica rocks share the shape of distant horizons (Fig. 4.6). Chthonic power, emerging from deep underground, would appear as manifest in many of these carved stones. In situ stone is rooted in the Earth, providing contact with those primordial powers and with the Pachamama. Stones that were unusual in shape or location seemed to have been especially venerated. If nature had marked out such a stone, it may have been viewed as something miraculous, an entry of the divine into the ordinary, a hierophany (Eliade 1972).

Carved huacas may also express mytho-historic traditions of the Incas, such as creation mythologies dealing with water and caves, animism, and shamanistic ritual. Ritual stairs are a dominant motif, perhaps expressing movement from below to the heavens, and are frequently associated with the three realms. Seats or thrones were often sculpted as part of these huacas, providing orientation to view mountains or solar events on the horizon. There is an association of stone huacas with water (Solomon and Urioste 1991) or other ceremonial liquids, joining origin myths and the practical needs of agriculture. Rituals may have involved the pouring of liquids into basins and channels (Fig. 4.7). Most astronomical huacas are near to streams or

Fig. 4.7 Ceremonial channel at the Sun Temple of Llactapata. (Author's photograph)

canals and appear to have been "given sentience" through camay. Most of the carved rock huacas surrounding Cusco are located in Chinchaysuyu and Antisuyu (Farrington 1992, p. 373).

Certain shrines were dedicated to the Sun with light and shadow effects highlighting times such as solstices and equinoxes. Huacas also may express explicit alignments to sunrise or sunset at specific times of the year. Features may designate vantage points for observing astronomical phenomena, guide the eye to solar horizon events or the rising of the Pleiades, and mark approximate dates by shadow or the casting of light onto altars in caves.

4.4 Huaca Maintenance and Worship

Huacas required many prayers and offerings and the responsibility for such was assigned to specific panacas and ayllus. Panacas were assigned for the care and conduct of rituals at huacas on certain ceques while ayllus were designated to care for others. Every ceque had a kin group assigned to manage its affairs (Zuidema 1964).

By design, huacas were self-sufficient with large tracts of land and animals to fulfill the needs of their attendants. These caretakers belonged to the assigned panaca or ayllu and many thousands were employed and supported in this manner (Bauer and Stanish 2001). This system served as the state's method of allocating territory and irrigation sources (Niles 1987).

Van de Guchte (1990) relates that a huaca was draped with textiles designating the specific panaca or ayllu responsible for its care. The attendants farmed the huacas' land, tended its flocks, and brought it special offerings on its specific day of celebration. Sacrifices were made at huacas, in part to support the attending panaca or ayllu, and these varied with the status of the shrine (Bauer 1998, p. 27; D'Altroy 2002, p. 167).

4.5 Ceque and Huaca Astronomy

Huacas can exhibit explicit cases of astronomical orientation and numerous examples have been identified. These often were alignments for the June solstice sunrise, while others pointed to the Sun at December solstice. Light-tubes or cave openings allowed altars to be illuminated at specific times and orientations guide the eye to the horizon on solar significant dates. Pillars were set on hills to calendrically mark the passage of the Sun on the horizon. I observed a ceremonial staircase to be illuminated by the Sun, as well as an animistic character created from light and shadow at sunrise. Both are among numerous examples of the astronomical passion of the Incas and their solar religion. Zuidema found through the chronicles and his field research that many ceques were astronomically oriented and pointed to the horizon to guide the eye to the Sun and stars.

4.6 Ceque System Controversy

Zuidema stated that the ceque system was utilized as a walk-through ritual calendar, with each huaca honored and worshipped on its own specific day of the year. He also proposed the alignment of certain huacas to have astronomical significance. Each huaca represented a day of the year and the specific number of huacas, 328, is important because it represents the lunar sidereal year, which is 12 × 27.3, the number of days in a sidereal lunar month. Zuidema suggests that the 37 days missing between this calendar and the tropical year may be the approximately 37 days that the Pleiades are not visible in the sky because of the proximity of the Sun. The period from May 2nd to June 9th also was one where there were no agricultural rituals to be timed and executed (Zuidema 1989a). The ceque calendar therefore breaks into 12 sidereal lunar months of 27 1/3 days with the 41 ceques also representing 41 weeks of 8 days each. The days of 12 month-like periods were counted by the huacas within each ceque group of three. In this way the number of days per month varied between 22 and 33 (Zuidema 1989a; Fig. 4.8).

The astronomical and calendrical basis of ceques has been rigorously argued by Zuidema and Aveni, but Bauer and Dearborn state that they find less evidence for

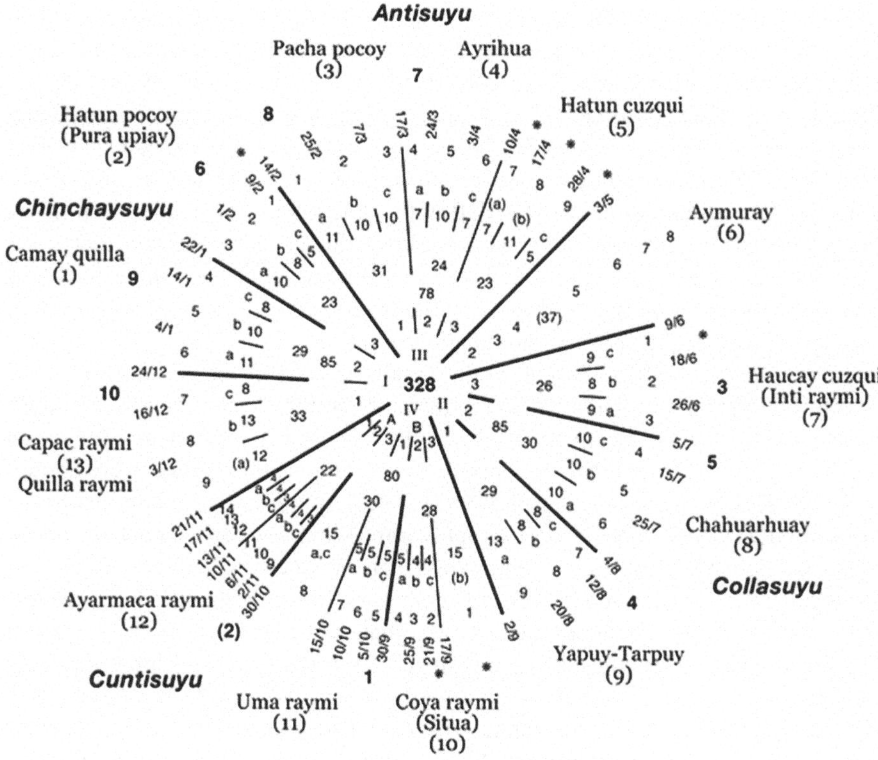

Fig. 4.8 A Cusco ceque-based calendar. (From Zuidema 2015)

ceque celestial sightlines (Bauer and Dearborn 1995, pp. 64–65; Zuidema 1977; Zuidema 1982a). Bauer and Dearborn (1995, p. 65) point out that the period of disappearance for the Pleiades can vary from year to year depending on the Moon's phase, the horizon altitude, and atmospheric conditions. They state that "For the Pleiades, a disappearance from the night sky lasting forty to fifty days is quite reasonable" (p. 130). Even if the specific reason for the 37-day gap remains elusive, the ceque calendar as described would have been a most efficient tool for the Incas to use in the management of Cusco and their empire.

Zuidema (1977, p. 251; 1982a, p. 204) argued that ceques formed straight lines. Others doubt that the paths of ceques were this direct (Fig. 4.2). Niles (1987, pp. 177–179), using descriptions by Cobo, place names, and Inca roads finds ceques not to be straight and instead that they followed paths with changing courses. One of Cobo's (1990 [1653]) huaca descriptions states "[Ch-9:4] The fourth *guaca* was a fountain named Pomacucho which was somewhat separated from the *ceque*" (p. 61). Still, this may not mean that ceques were not intended to be straight as the related huacas may have straddled a conceptually straight sightline. Niles maintains that while ceques never crossed one another, they did not proceed straight from the Coricancha to the horizon. Bauer (1998) mentions that "Zuidema and Aveni do not state that all of the ceques were straight," but also adds "it is the specific locations of the huacas that define the course of the lines and not vice versa" (p. 11). Bauer agrees with Niles and says that certain huacas on certain ceques varied considerably to the left or right of a straight line. They assert that this variance was great enough to have made it unlikely that ceques could have formed sightlines to guide the eye to celestial horizon events (Bauer and Dearborn 1995, pp. 132–133; Niles 1987, pp. 177–179). Zuidema (1977) stated that "While keeping the concept of ceque as a straight line, huacas of one ceque could be found to the right or left with some margin of freedom" (p. 251). Aveni (1981d) argues that his research with Zuidema demonstrates that some ceques are straight, one represented an astronomical sight line, three locations were used for observations, and two sightlines crossed ceques employing certain huacas on those ceques for astronomical observations.

Niles (1987, p. 178) says that while certain huacas played a role in Cusco's calendrical system it is difficult to argue that astronomical orientation was a determining factor in the placement of all shrines and ceques. Bauer and Dearborn (1995, pp. 130–133) feel that evidence does not support that ceques were associated with specific stars. Farrington (1992) states "Importantly, in contrast to the work of Niles (1987: 178), Zuidema and Aveni (Aveni 1981a, p. 313) have established by field verification and toponymy that the lines are straight, even over distances up to 20 km; about thirty of them have been mapped with reasonable precision" (p. 370). Zuidema (2007) asserts that the "ceque system does account well for the Cuzco calendar as reconstructed from the solar and lunar data and corroborated by the information on ritual movements" (p. 279). Cobo (1990 [1653], pp. 60–61) listed the tenth huaca on Chinchasuyu's eighth ceque as *Catachillay*, a name also related to the Pleiades. The seventh huaca on the same ceque was called *Sucanca*. The pillars of Sucanca, as viewed from the Coricancha, indicated the position of the Pleiades' last heliacal set about April 15th (Zuidema 1977). The evidence from the chronicle of Cobo, when taken with observations in the field, supports that this ceque had an astronomical alignment from the Coricancha to the horizon.

While Bauer has said that ceques could not have been straight because of their huacas having been situated in zigzag-like patterns, Zuidema's description could still work. Zuidema (1977, p. 251) states that ceques were sightlines to points on the horizon and not necessarily direct connectors of huacas. He adds that the huacas associated with a particular ceque could fall to the left or right of this sightline. Such a case is in contrast with Bauer's argument that ceques followed zigzagged paths connecting their corresponding huacas and could allow further support of Zuidema's model of ceque sightlines to the horizon. Bauer's statement that ceques did not cross each other (Bauer 1998, p. 11) works with this interpretation of Zuidema's concept. The ceques in Zuidema's model do not cross, even if their haucas lie not on them but only in close proximity. Though many of these huacas did not fall exactly on the straight lines that gave them order, the actual sightlines between them were important factors in their care and worship. "A ceque probably was something more than a loose line drawn from one accidental huaca location to the next" (Zuidema 2002, p. 593).

Bauer and Dearborn agree with Zuidema that the Incas appeared to exhibit definite interest in the Pleiades (Bauer and Dearborn 1995, p. 124; Fig. 6.17). This might have been as part of an Andean methodology for agricultural forecasting by observing the relative brilliance of this star group following its heliacal rise (Orlove et al. 2000). Zuidema and Aveni identified a corridor in the Coricancha they said that opens though the southwest wall of the of the temple's courtyard to face in the direction of the rise of the Pleiades, c. 1500 CE (Zuidema, 1982a, pp. 212–214). Such an orientation to guide the eye seems plausible for a society observing the Pleiades due to concern about future harvests, but Ziólkowski and Kościuk (2018) found less evidence to support such alignments during their research. The Sun temple at Llactapata exhibits a similar alignment, which is further discussed in Sect. 9.4.

4.7 Summary

The Incas ordered their religion and society by the ceques and huacas of Cusco. The ceque system exhibits certain similarities with the quipu, a woven mnemonic aid whose many strands and knots radiate from a central focal point. Ceques may as well have been used to organize shrines of worship at other key locations within the empire, one potential example of such being identified in the vicinity of Machu Picchu. The specific purpose and alignment intended for Cusco's ceques remain a matter of dispute.

Huacas were shrines venerated due to the great spiritual entities within them and at least 328 were known to surround Cusco at the time of the Spanish invasion. There likely were many more, especially those intended for personal worship not sponsored by the state. Ceques were assigned to specific panacas and ayllus, who in turn were responsible for the care and worship of the huacas found along them. Catholic priests set out to extirpate everything regarding the indigenous religion and largely succeeded, exceptions being the many carved rock huacas they could not destroy.

Chapter 5
Archaeoastronomy

Fig. 5.1 Entrance to the southeast cave at Lacco. (Pencil by Jessica Gullberg)

© Springer Nature Switzerland AG 2020

S. R. Gullberg, *Astronomy of the Inca Empire*, Historical & Cultural Astronomy,
https://doi.org/10.1007/978-3-030-48366-1_5

Archaeoastronomy is sometimes said to be the study of the anthropology of astronomy. Researchers seek to establish cultural patterns for alignments associated with heavenly bodies that are easily observable to the naked eye. Celestial orientations appear in the architectural constructs of many societies, implying there was great ancient interest and dependence upon the Sun, Moon, planets, and stars to regulate or otherwise influence many functions of everyday life. Astronomical positioning was far more than just a means of telling time, however, as celestial occurrences appear to have been intertwined with multiple aspects of society and culture. In the past, classic archaeoastronomy concerned itself more with alignments as it left cultural interpretations to ethnoastronomers, but today lines have become much more blurred as archaeoastronomers, too, seek answers to why and for what purpose the orientations they discover exist.

Certain research has been aided with written records left by societies of their astronomical activities, such as the Babylonians, Egyptians, and Maya. Others like the Inca had no system of writing, thus complicating efforts to place their celestial practices into proper context. Archaeoastronomy relies heavily upon interpretations of the azimuths and elevations of sightlines crafted into constructs such as shrines and temples. It also documents effects of light and shadow.

This chapter explores interests, methods, parameters, and celestial occurrences that have been observed and practiced, some for millennia. Archaeoastronomical methods are examined that relate to ancient societies and especially to the Incas with regard to their versions of astronomy and cosmology.

5.1 The Celestial Sphere

People of all civilizations throughout history have observed similar celestial motions. Their interpretations, explanations, and uses for them are what differ. Most present-day astronomers have very little time to contemplate repetitive patterns and positions of heavenly bodies as they rise or set on the horizon. Many ancient cultures studied such events closely, however, and were very much experts on what can be called "horizon astronomy."

These astral phenomena result from the diurnal motion of the Earth's rotation as well as the annual travel of the planet in its orbit about the Sun. In the Southern Hemisphere the Earth's apparent motion is from left to right, or clockwise, when looking north toward the equator. As a result the stars, Sun, Moon, and planets all rise from the east to the right and set to the left in the west. They appear to be on a celestial sphere, equidistant in the sky, and travel upon it in their daily motions. Predictable movement upon this sphere caught the attention of many an ancient observer and gave rise to the surprisingly sophisticated astronomies that we discover today.

5.2 Motions of the Heavens

While all looked essentially upon the same Sun, Moon, planets, Milky Way, and certain stars, ancient observers in various parts of the world envisioned many different things. The figures imagined in the sky by one civilization did not necessarily match the constellations devised by another. These individual reflections gave rise to a number of different cosmologies that were influenced by the orientation of the view of the heavens offered by the geographic location of the civilization in question (Urton 1981a). We often view astronomical occurrences from our own perspective and therefore need to make a conscious effort to see the sky from the eyes of those we study (Aveni 1981c).

Mid-Latitudes

In the mid-latitudes, bodies traveling on or near the ecliptic appear to rise in the east, cross the sky in an arc, and set on the western horizon (Fig. 5.2 – right). If viewed from the Southern Hemisphere looking north, these bodies travel right to left, and in the Northern Hemisphere looking south, from left to right. The altitude of the arc of travel may vary throughout the year and is dependent upon the latitude of the observer and the position of the Earth in its annual orbit with regard to the present 23.439° (23.44°) tilt of the Earth's axis. Annual variations in the height of this arc are responsible for the consistently changing solar rising and setting points on the horizon that inspired many of the astronomical practices of ancient civilizations.

Stars, as well as the Sun and Moon, travel east to west on their own respective arcs. In the Southern Hemisphere, when looking to the south, those stars in proximity to the south celestial pole are seen to travel in circles around the pole due to the Earth's rotation about its axis. These stars are called *circumpolar* and the number of them that can be viewed as such, always above the horizon, is dependent upon the latitude of the observer. As viewed from Cusco, these would be the stars whose circles' lower limbs lie within 13° of the southern horizon.

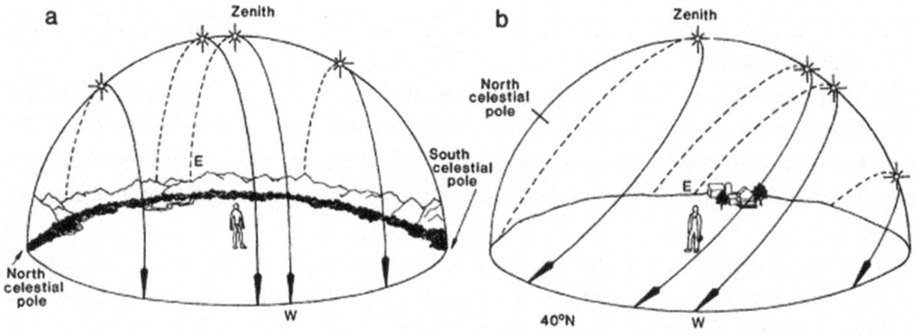

Fig. 5.2 Paths of apparent celestial travel in the equatorial latitudes (**a**) and the mid-latitudes (**b**). Motion in the tropics is vertical while that in the temperate zones is oblique and circular. (From Aveni 1981c, p. 163)

Polar Latitudes

In the polar regions, more specifically at 90° north or south at the celestial poles, all visible stars appear as circumpolar and trace concentric circles in the sky. The Sun, Moon, and planets cross the sky on or near the arc of the ecliptic, which is at an altitude dependent upon the seasonal variation of the orientation of the Earth in its orbit.

As viewed from the poles the ecliptic can cross the sky no higher than 23.44° above the horizon during the hemispherical summer. For half the year during the winter, the ecliptic is below the horizon, and those bodies traveling on or near its path cannot be seen. These orientations result in days without sunset or sunrise. The Sun will rise once at the associated hemispherical spring equinox and will not set again until the corresponding equinox of the fall. Latitudes greater than 66.56° experience lesser degrees of total day or night, accordingly.

Equatorial Latitudes

Observers on the equator see stars rise perpendicular to the eastern horizon, climb directly up and across the sky, and set perpendicularly on the western horizon (Fig. 5.2 – left). The lines traced by each star are parallel to one another.

On the equinoxes the ecliptic is oriented with the celestial equator so that bodies traveling on or near it rise and set straight up from the east and down in the west. During the rest of the year the Sun travels as much as 23.44° north and south of the celestial equator.

Latitude of Cusco, Peru

Cusco lies in the Andes at 13.532° south latitude. In such a tropical region the Incas experienced apparent celestial motions between those of the equator and middle latitudes. In Cusco, when facing north, the Sun and other ecliptic bodies will rise in the east on the right, cross the sky in arcs to the north, and set on the left in the west (Fig. 5.3). The region surrounding Cusco experiences a Sun in June that rises in the northeast, crosses the sky on a relatively lower arc, and correspondingly sets in the northwest. In December the solstice Sun rises in the southeast, follows a higher path, and sets again in the southwest.

The Incas and the civilizations preceding them apparently observed and carefully noted the repetitive and predictable patterns of the horizon positions of sunrises and sunsets. They used this knowledge to develop a calendrical system for crop management and denoting recurring religious festivals. The Incas built pillars on certain smooth horizons to predict and mark the positions of solar events at key times of the year for public displays affirming the status of the ruling elite. In other locations they may have made comparable observations with rugged natural geographic features.

Inca knowledge of solar alignments extended into their architecture and shrines, where many have been found to be oriented for purposeful effects of light and shadow. Certain temples were constructed with regard for solar and Pleiades horizon events, as well as were huacas designed for special ceremony and purpose. Several of these shrines have features that are illuminated by the Sun or Moon at significant times.

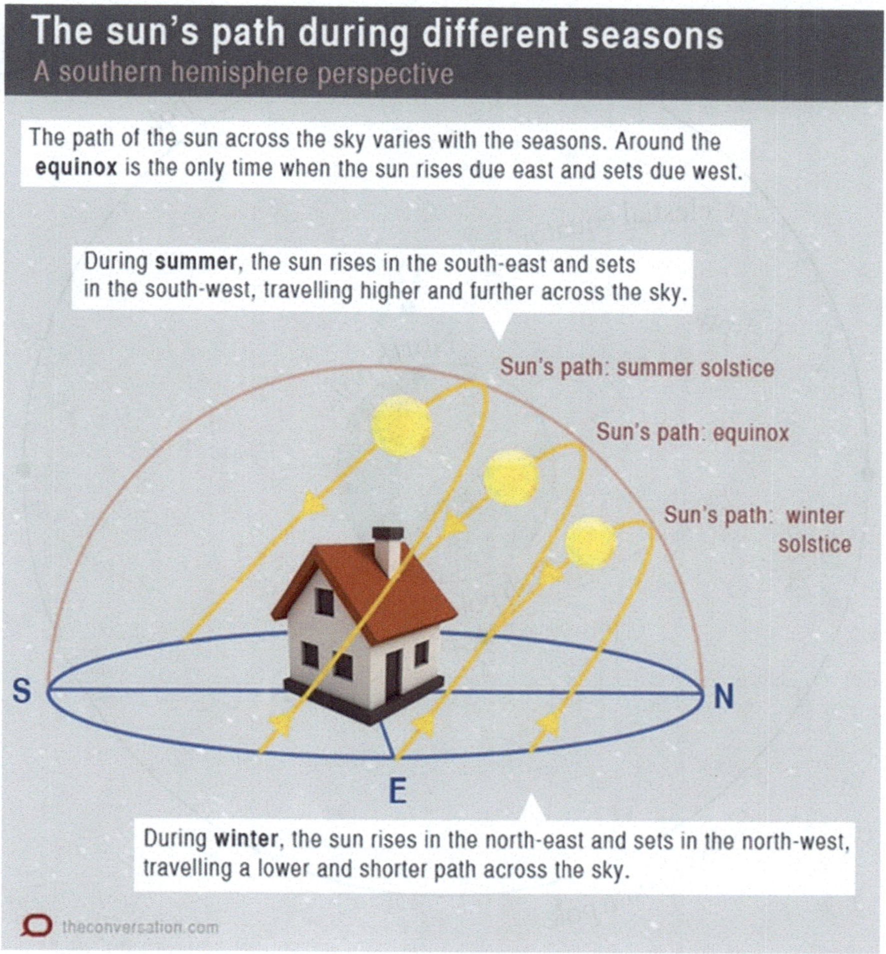

Fig. 5.3 The paths of the Sun across the sky in the southern hemisphere. (Courtesy of *The Conversation*)

5.3 Solstices and Equinoxes

The moment that the Sun on its ecliptic path crosses the celestial equator is called an *equinox*. This occurs twice each year on or about March 21st and September 22nd and on those dates the crossing results in days with equal periods of sunlight and darkness. The longest days and nights occur annually when the Sun reaches the northernmost and southernmost points in its ecliptic travel (23.44° north or south) on or about June 21st and December 21st. These are the June and December solstices, and are commonly known as the summer and winter solstices at opposing times in the respective hemispheres (Fig. 5.4).

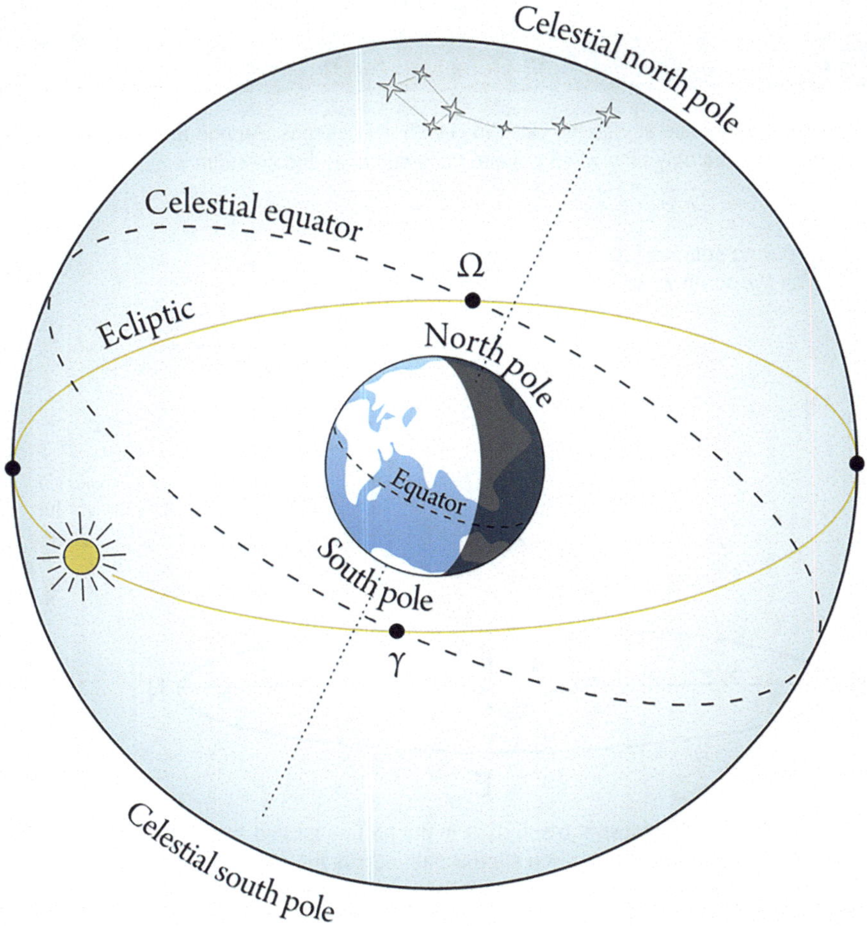

Fig. 5.4 The solstices and equinoxes with the celestial equator. (Joshua Cesa, Public domain / https://upload.wikimedia.org/wikipedia/commons/4/47/Celestial_equator_and_ecliptic.svg)

Awareness in ancient cultures of the existence of solstices and equinoxes likely came through observations made over many years. Persons of intellectual curiosity, such as priests, watched the recurring motions of the Sun, Moon, stars, and planets with wonder and noted the patterns that they made. Given enough time they observed dependable cycles and devised methods to both predict and utilize them. Inca awareness of the equinoxes is mentioned in Spanish chronicles (Garcilaso 1961[1609], p. 72), but conclusive evidence is elusive. Additionally, in his narrative, Garcilaso confuses equinoxes with zenith passage (Sect. 5.5). Solstices are easy to identify when the daily sunrise position ceases to move on the horizon, but horizon observations at times of the equinoxes at the latitude of Cusco are less precise due to the 25 arc-minute per day rate change in the position of the rising Sun (Dearborn and White 1989, p. 468).

The Incas learned the cycles of solstices and equinoxes and used this knowledge as a key component of their annual crop management activities. Zuidema (1981b) and Aveni (1981a) describe pillars, no longer extant, on the horizon of Cusco that were mentioned by chroniclers following the Spanish conquest. These are discussed in Sect. 6.8.

5.4 Cardinal Directions

Cardinal directions were known to the ancients and could possibly have been determined by using the shadow plots of gnomons. On a curve traced by points plotted of the tip of a gnomon's shadow throughout the day, the point at which the shadow was closest to the gnomon indicated true south in the Southern Hemisphere (Evans 1998). North is opposite south and the directions of east and west are perpendicular to this line.

A circle that is traced through the shadow plot at two points, with its center at the gnomon, can be used for more precise measurement. Lines may be drawn from the intersecting points to the gnomon thus forming an angle that, when bisected, indicates through the shortest shadow to the south (Evans 1998). While within their technological capabilities, there is no conclusive evidence that the Incas used these procedures.

5.5 Zenith and Anti-Zenith Sun

As the Sun travels along the ecliptic, twice per year it passes directly above any point lying between the Tropic of Cancer and the Tropic of Capricorn (Fig. 5.5). Locations precisely situated on either tropic experience the zenith Sun only once on the day of the associated solstice. All sites between the latitudes of N23.44° and S23.44° observe times without shadow when the Sun passes directly overhead. This includes all of Peru and the greater part of the former Inca Empire. Exposed to the phenomenon of zenith passage, civilizations between the tropics exhibited aspects of their cosmologies that varied somewhat from those further to the north or south (Urton 1981b).

Actual dates for the zenith Sun relate to the latitude of the observer. In the vicinity of Cusco the two dates for zenith passage occur on February 13th and on October 30th (Zuidema 1981b, pp. 322–323). Zenith passage can be observed when a vertical gnomon casts no shadow at local noon, and on those dates the Sun rises on the Cusco horizon at approximately 103.5° and sets at about 256.5°. This is discussed further in Sect. 6.6.

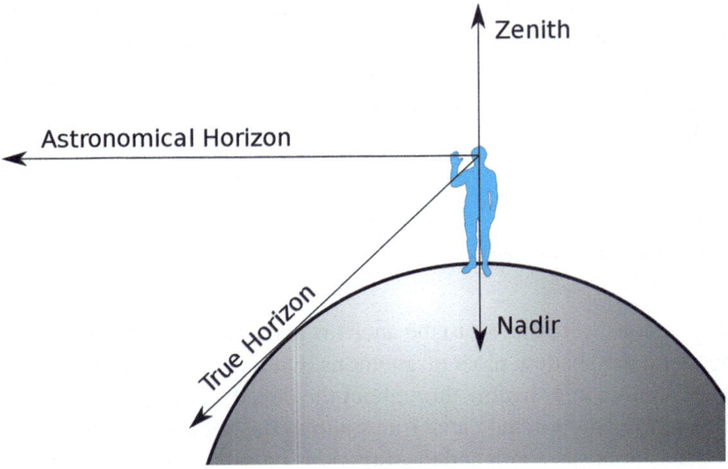

Fig. 5.5 At zenith the Sun is directly overhead; anti-zenith is when the Sun is at nadir. (Credit: Wikigraphist DTR, Graphics Lab, Wikimedia Commons)

Zuidema (1981b) suggests that the Incas also observed days of anti-zenith, or nadir, passage of the Sun (Fig. 5.5), and states:

> The only way in which the Incas could have determined the dates of August 18 and April 26 would be to note the days when the sun goes through the zenith. Reversing the direction of sunrise on those days would have given them the days of antizenith (= nadir) sunset. (p. 322)

Anti-zenith passage occurs in Cusco on April 26th or 27th and August 18th or 19th, with sunrise on these days occurring at 076.5° and sunset at 283.5°. These dates closely correlate with Inca festivals of maize planting and harvest. Zuidema (1981b) relates that the Incas used the zenith sunrise and anti-zenith sunset for such purposes.

Garcilaso (1961 [1609]) gives an account related to a method of observing zenith passage. In it he describes priests as making daily observations of a shadow cast by a column as the time of the equinox approached. A large circle was drawn with an east-west line through the column at its center. They observed the approach of the equinox with this shadow and, according to Garcilaso, when the shadow was "reduced by half, from sunrise to sunset" (p. 72) and the column cast no shadow at noon, they knew the equinox had arrived. Garcilaso's wording, however, exhibits confusion between the observance of the equinox and that of the zenith Sun. Inca interest in the anti-zenith passage of the Sun is further discussed in Sect. 6.6.

5.6 Horizon Astronomy

As the Earth revolves its axial tilt causes a daily change in the horizon positions of sunrises and sunsets. Ancient civilizations used this knowledge to develop systems of timekeeping to regulate planting and harvesting, as well as recurring religious festivals and other civic events.

Natural features were identified to mark the positions of the Sun on days of solstices and equinoxes. Alignments of buildings and shrines were also crafted to assist in identifying these important dates. The Incas erected pillars on the relatively smooth horizons of Cusco to assist in observing these solar phenomena, and possibly as well for zenith and anti-zenith sunrises and sunsets (Bauer and Dearborn 1995). Observations were made from key points in Cusco to identify annual horizon positions. It has been suggested that the Incas may even have erected sufficient pillars to denote "months" throughout the solar year. For these and other events the Incas made valuable use of the predictable and reliable patterns of the Sun on the horizon as a calendar to regulate their daily lives.

The region surrounding Cusco experiences June solstice sunrise at approximately 064° and has a corresponding sunset at 294°. In December the solstice Sun rises at 114° and sets again at 244°. The 50° span of the horizon between the two solsticial points of sunrise and the two solsticial points of sunset was of particular interest to the Incas. Many huacas have been found to be oriented to these key horizon events.

5.7 Field Research

In addition to outlining the fundamental principles of archaeoastronomy, it is important to introduce certain factors related to collection of data in field research. This section will discuss tools, trigonometry, and atmospheric considerations.

Archaeoastronomical Implements

Tools typical in archaeoastronomy include theodolites, GPS receivers, sighting compasses, inclinometers, digital cameras, tape measures, computers, and astronomical software. The theodolite, or surveyor's transit, measures angles very precisely, to the arcsecond, and is used to verify alignments requiring the utmost accuracy (Fig. 5.6). GPS receivers can provide coordinates of latitude and longitude, as well as elevation above sea level for each research object or site. Certain values may be determined and used in conjunction with appropriate formulae found in Sect. 5.7. A sighting compass is useful for measuring azimuths and alignments in far less time than that required by a theodolite (Fig. 5.7). An inclinometer is used to determine angles to objects above or below the level horizon (Fig. 5.7). Digital photographs are useful for documenting every aspect of a research site, including any evidence of effects of light and shadow. Tape measures are used to record dimensions of certain object(s) under study. Computers and astronomical software provide insight into skies that were viewed by ancient societies and are useful in predicting alignments and celestial events. *Google Earth* is also a valuable tool. This program makes available world-wide satellite imagery in formats that allow

Fig. 5.6 Mike Zawaski using a theodolite at Q'espiwanka. (Author's photograph)

visual evaluation of potential alignments. Other sophisticated means such as Stellarium are available as well.

Spherical Trigonometry

It often is useful to be able to predict the positions of horizon events. Movement of astral bodies on the celestial sphere is defined mathematically and can be precisely calculated with spherical trigonometry, a foundation of archaeoastronomical research. Formulas may be used to calculate angles that specify horizon points of celestial bodies, and also to verify field data taken by sighting compass, GPS, and theodolite.

In horizon astronomy research, when the inclination (angular altitude) of the horizon at or above level is known by measurement with such as an inclinometer, the latitude of the site in question is taken from a chart or GPS measurement, and the declination of the Sun is derived from the nautical almanac, the actual position where the Sun will rise on an elevated horizon can be determined.

These values can be determined with software, but the relevant formulae of spherical trigonometry are listed in Appendix A.6.

Fig. 5.7 Sighting Compass and Inclinometer. (Author's photograph)

Temporal and Atmospheric Phenomena

Several factors can affect the precision of archaeoastronomical field work, such as precession and obliquity, refraction, extinction, and horizon deviation. When appropriate, these must be taken into consideration to achieve precise results.

Precession and Obliquity

Precession of the equinoxes occurs due to the slow wobble of the Earth's axis. This nearly 26,000-year cycle causes the celestial poles and equator to migrate when compared against the background of the stars (Fig. 5.8). Right ascension and declination change over time as a result and the difference in declination leads to new rising and setting azimuths for stellar objects. Ancient astronomical orientations that might not be obvious today can be determined with declinations adjusted for the time period when a temple or shrine was designed.

Bodies within our Solar System traveling on or near the ecliptic are not affected by precession. The obliquity of the ecliptic, the inclination of the Earth's equator to the plane of the ecliptic, is presently 23.439°. This decreases at a rate of about 47

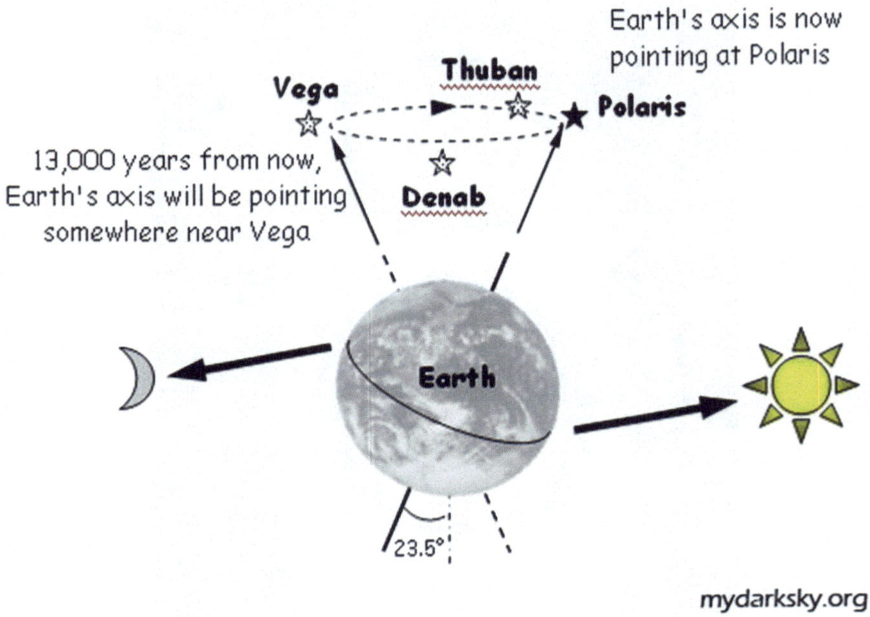

Fig. 5.8 Precession of the equinoxes. (Courtesy of mydarksky.org / https://mydarksky.org/2008/10/14/the-earths-wobble-precession/)

arcseconds (0.013°) every 100 years and therefore is a consideration for celestial measurements regarding ancient structures (Fig. 5.9).

Refraction
Celestial bodies appear higher in the sky than they actually are due to atmospheric refraction. This is the result of light being slightly deflected due to differences in atmospheric density and has a greater effect near the horizon (Schaefer 1993). The azimuth of an object is shifted due to refraction, an effect which is more pronounced at higher latitudes with smaller incident angles to the horizon than in the tropics. The effect of refraction becomes negligible at altitudes 20° or more above the horizon. Since refraction is sensitive to the atmospheric profile integrated along the line of sight and since that profile is influenced by local parameters, corrections are increasingly uncertain for objects near the horizon. The effect becomes pronounced at higher latitudes and is generally insignificant in the tropics (Fig. 5.10).

Extinction
The absorption and scattering of photons when colliding with particles or air molecules is known as atmospheric extinction. Atmospheric water, carbon dioxide and oxygen are primary absorbers. *Rayleigh Scattering* occurs with air molecules smaller than the wavelength of the photons in question (Fig. 5.11). *Mie Scattering* results similarly with photons striking small atmospheric particles of similar wavelengths (Bely 2003). Extinction is not an issue when dealing with the Sun and

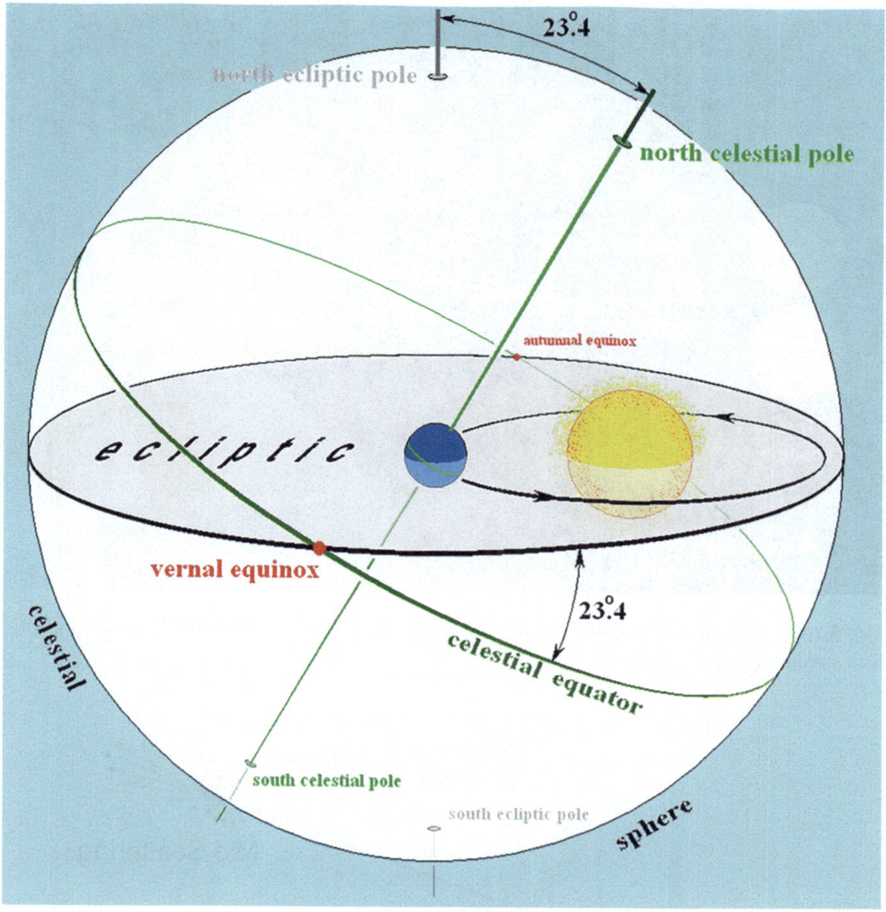

Fig. 5.9 Obliquity of the Ecliptic. (Tfr000 (talk) 01:59, 15 March 2012 (UTC) / CC BY-SA / https://creativecommons.org/licenses/by-sa/3.0)

Moon, but becomes significant when observing faint stars near the horizon, such as the Pleiades (Schaefer 1986).

Horizon Deviation

Horizon deviation occurs when the visible horizon differs from the astronomical horizon. This happens commonly in areas of mountainous terrain and is most pronounced at higher latitudes where the arc of travel for a specific body can yield a significantly different point for rising or setting than that given for the astronomical horizon. Examples of this are shown in Table 5.1. Values including horizon deviation as derived with formula (A.8) in Appendix A.6 are listed in Appendix A.5. Adjustments for horizon deviation are needed frequently in archaeoastronomical research.

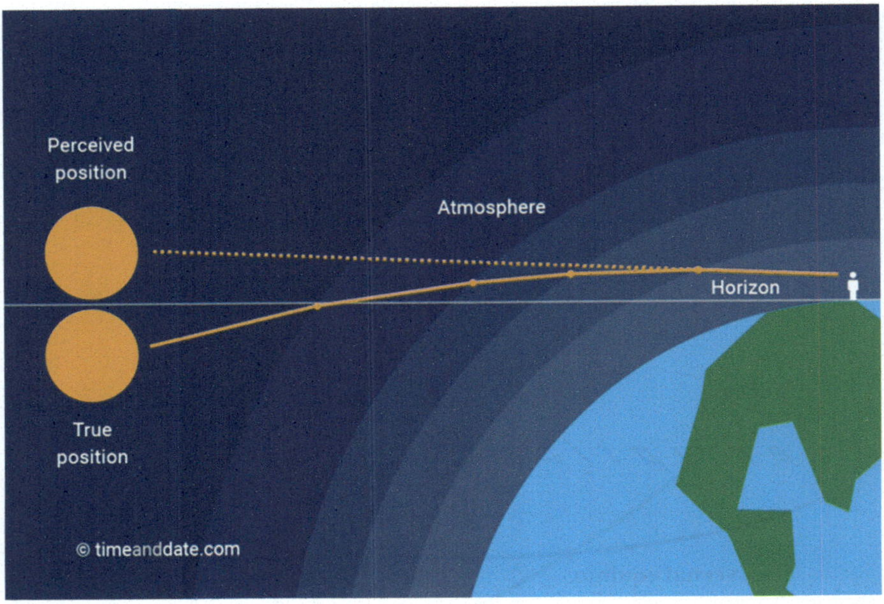

Fig. 5.10 Atmospheric refraction. (Courtesy of TimeAndDate.com / https://www.timeanddate.com/astronomy/refraction.html)

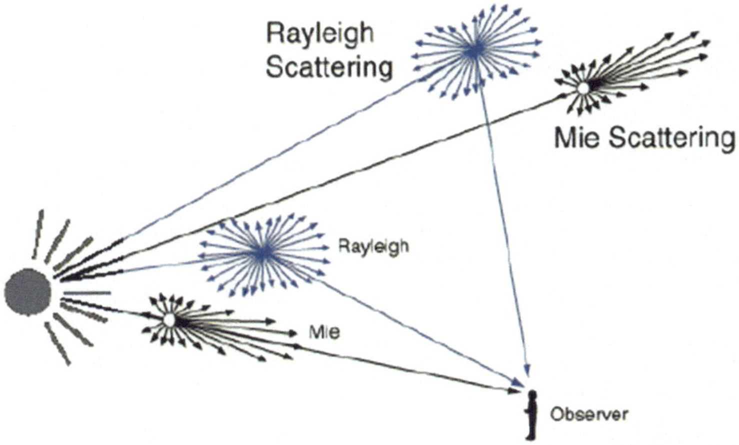

Fig. 5.11 Rayleigh and Mie Scattering. (From Kment et al. 2006)

Related Research Considerations

Precession is not a concern for solar observations, but for measurements involving stars such as the Pleiades it must be taken into consideration. The obliquity of the ecliptic has changed 0.0649° since the time of the Incas, resulting in a difference in azimuth of 4′ at a latitude of 13° and is therefore negligible. While refraction and extinction can be important when examining the rising and setting of stars, they also

Table 5.1 Horizon Azimuth Shift for Altitude. (Author's table)

Altitude above astronomical horizon	Latitude 13.5°	Latitude 40°
1°	0.22°	1.04°
2°	0.49°	1.99°
4°	1.07°	3.84°
6°	1.69°	5.63°
8°	2.35°	7.37°
10°	3.04°	8.06°

are not a concern for research involving the Sun. The rising and setting of all celestial bodies at the S13.5° latitude in Cusco, the Sacred Valley, and Machu Picchu must take horizon deviation into account.

5.8 Summary

Study of the constructs of ancient societies in an astronomical context requires both careful measurement and the consideration of pertinent phenomena. Paths of heavenly bodies on the celestial sphere vary with latitude, and throughout Cusco, the Sacred Valley, and Machu Picchu motions observed are typical of those found in the tropics. Solar paths arc across the northern sky, both rising and setting with pronounced angles of incidence with the horizon. This latitude also experiences solar zenith on February 13th or 14th and October 30th each year.

Solstices and equinoxes of the Sun were of primary interest in a number of ancient cultures and sometimes were used, including by the Incas, as a calendar to regulate crop management and recurring religious festivals. The Incas were found to at times have erected pillars to assist them in recognizing these key astronomical events when natural horizon features were insufficient.

Cardinal directions were important to certain ancient civilizations and could be determined through the use of gnomons designed for this purpose. The precise moment of the zenith Sun could also be observed through the absence of shadows from vertical devices.

Horizon astronomy is the observation of solar, lunar, planetary, and stellar rising and setting events on the horizon. Numerous ancient civilizations, including the Incas, practiced horizon astronomy in an effort to better regulate their society on an annual basis.

Various instruments and spherical trigonometry are valuable tools in archaeoastronomical research. Implements include theodolites, GPS receivers, sighting compasses, inclinometers, digital cameras, tape measures, computers, and astronomical software. Mathematical formulae and/or computer software may be used to accurately depict, predict, or verify the horizon positions of various astronomical events. Researchers must be cognizant of and adapt for, as necessary, the effects of precession, obliquity, refraction, absorption, scattering, and the degree of deviation of the skyline from the astronomical horizon. Of these factors, horizon deviation is the principle concern for solar research in the Andes.

Chapter 6
Inca Astronomy and Cosmology

Fig. 6.1 An altar within an inner chamber of an Inca cave at Lacco called the Temple of the Moon that is illuminated around local noon at and near the time of the zenith Sun. (Watercolor by Jessica Gullberg)

© Springer Nature Switzerland AG 2020

S. R. Gullberg, *Astronomy of the Inca Empire*, Historical & Cultural Astronomy,
https://doi.org/10.1007/978-3-030-48366-1_6

Astronomy in the Andes was well developed by the time the Spaniards arrived. This was due in large to the accumulation of knowledge through observations made by the many civilizations preceding the Incas. Astronomy was not simply observing and understanding celestial movement, however, as it was integrally woven into the very fabric of Andean existence throughout myth, cosmology, and culture, playing an important role in daily life.

The Incas were a Sun-worshipping people and their emperor was said to be "the son of the Sun." Their cosmology begins with the primordial rising of the Sun, and also that of the Moon. In their astronomy they were aware of many stars and planets and paid particular attention to the Milky Way and the Pleiades.

In a practical sense, this knowledge was put to work via horizon astronomy, as the Incas marked the passage of sunrises and sunsets on their horizons in order to keep time for agriculture and religion. Ultimately, celestial alignments were integrated into their temples and huacas, as well as with other constructs such as solar pillars built more specifically for astronomical purposes. The intriguing astronomy found in Inca huacas is much better understood when placed in the context of Andean astronomy as a whole.

6.1 A Complex Astronomy

Pachacuti Inca Yupanqui was the ninth emperor of the Incas and succeeded his father, Viracoca Inca, following a legendary battle with the Chanca in 1438 where it is said that Pachacuti led his people to victory while his father fled from Cusco. This pivotal conflict took place a mere two centuries after the founding Incas first entered the Cusco valley. The Spaniards began their invasion of the Incas' domain nine decades later and thought them to be far too undeveloped to have any understanding of the heavens (Aveni 1981a).

The conquistadors and priests were far from correct; in fact, the Incas' astronomical traditions began long before the beginnings of their empire. As the Incas conquered the Andes and assimilated tribe after tribe, they also took possession of cultural traditions and collective knowledge. Descriptions of recurring astronomical patterns realized through centuries of celestial observations made by successive civilizations were part of this adopted store of intelligence. Inca emperors refined this cosmology to best support the legitimacy of their power, but their concepts had long before begun with others such as the Huari, Nasca, and Chavin (Aveni 1981b). By the time of the 1532 invasion, Inca astronomy had developed to where it was considerably complex.

Astronomy in Cusco was interwoven into religion and everyday life, and the Incas believed heavenly activity to be directly connected with that on the Earth (Urton 1981a). Constellations were established with heliacal risings that foretold of coming terrestrial events. State efforts to promote the legitimacy of both solar worship and the emperor's power required an intimate knowledge of the observable phenomena of the Sun. These demonstrations were both for the benefit of the Inca populace and for newly conquered tribes. Successful crop management was dependent upon optimal times for planting and harvest and the Incas regulated this by horizon positions of the Sun. The spiritual world of the empire was also entwined with the heavens, not only by the gods of the constellations, but also through an intimate cosmology emanating from the Milky Way. Inca emperors had temples and huacas built with orientations for solar worship, festivals, and agriculture. Their ancient society was continually aware of happenings in the heavens.

Many Inca traditions survived the Spanish conquest through the residents of remote Andean villages. While priests made fervent efforts to convert the country to Christianity, many practices of indigenous culture and religion, including those astronomical, survived in these distant communities.

In the mid 1970s, Gary Urton (1981a) made a breakthrough in understanding Andean astronomy and cosmology that may hold added insight for Inca astronomy. In fieldwork for his doctoral dissertation, he lived for extensive periods amongst the villagers of Misminay in rural Peru, in a society where Quechua is still the daily language. Many of the ancient traditions remain, and Urton eventually was able to earn the trust of the residents. Gradually they shared cosmological philosophies that had been passed down through the ages since the time of the Incas (Urton 1981a).

Urton learned of their system of time reckoning using the Sun by day and the Moon or stars by night, and also the ancient names of 40 stars and constellations. The planting and harvesting of various crops were regulated month by month on an annual schedule that included the use of cycles of the Moon. Sunrises were utilized for times of planting and related horizon positions had been carefully determined to minimize crop failure (Urton 1981a).

Villagers also described the sky as being quadrapartitioned into four parts, as defined by two alternating intercardinal axes of the Milky Way. The community below mirrored this quadripartition and an intricate cosmology interrelated the two (Urton 1981a) (Fig. 6.2).

Urton (1981a) determined that:

> It seems reasonable to conclude that the concept of the center/planting sun (the sun from August to late October) which is found today in Misminay is a "descendant" of the Incaic practice of calculating the time of planting maize by means of solar pillars. Even though the

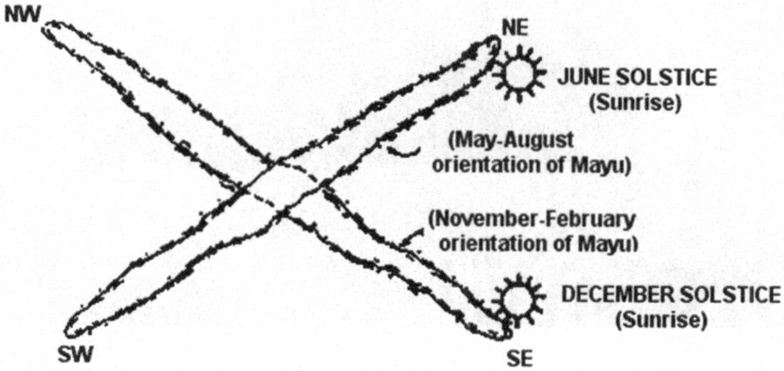

Fig. 6.2 Inca Quadripartition of the Sky. (From Urton 1981a)

> solar pillars (sucancas) which were in Cuzco during the time of the Incas were systematically destroyed during the "extirpation of idoltries," we may see the virtual embodiment of them in the saints' days (i.e., four saints' days = four solar pillars) and in such concepts as a planting and harvesting sun. (p. 77)

While there is no written evidence, much of the Incas' astronomy survived and exists through the memories and practices of contemporary Andean people.

The Incas' most significant Temple of the Sun, the Coricancha of Cusco (Fig. 6.3), was of preeminent importance as a focal point of state astronomy. The Coricancha's west wall is aligned with approximately the same azimuth as that for the point on the horizon of the rise of the Pleiades in June (Zuidema 1982a). Additionally, the December solstice sunset was observed from the Coricancha to two pillars on the horizon at Chinchincalla (Aveni 1981a, p. 308).

> [Cu 13:3] The third, Chinchincalla, is a large hill where there were two markers; when the sun reached them, it was time to plant. (Cobo 1990 [1653], p. 83)

In a nation whose religion and power centered on the Sun, astronomy and its related demonstrations were key. Maintaining public festivals as precise annual recurrences was an important part of exerting this influence over the masses (Bauer and Dearborn 1995). Elaborate ceremonies for both the elite and commoners were carefully crafted and celebrated, especially at the time of the solstices.

The Incas knew that the Sun took a year to complete its cycle (Hemming and Ranney 1982) and their view of this journey included its path along the horizon. The exact solstices and equinoxes of the Sun were well familiar, as were the days of

Fig. 6.3 The Coricancha. (Diego Delso / CC BY-SA / https://creativecommons.org/licenses/by-sa/4.0)

zenith and anti-zenith Suns. The Sun was used to determine exact days and hours for planting (Hemming and Ranney 1982) and became part of religious ceremony through carefully designed effects of light and shadow. In a society that preceded Copernicus, Kepler, and Galileo, Inca astronomy was relatively well developed.

6.2 Sun Worship

Founders of the Inca Empire sought to legitimize their power through state-mandated worship of the Sun. They created an elaborate cosmology centered on solar relationships and developed ceremonies and rituals to reinforce and perpetuate these state-serving myths.

The empire maintained the Sun to be the Inca's preeminent ancestor and the emperor to be the son of the Sun. Therefore, the power base and the legitimacy of the Inca royals depended upon the Sun as their source. The Sun was also revered for its life-giving role in agriculture.

Rituals and ceremonial travel to pilgrimage centers were both part of an elaborate system developed to establish and maintain these solar relationships and state ideology (Bauer and Stanish 2001; Sect. 3.16). The emperor Pachacuti ordered all his subjects to worship the Sun and constructed solar temples throughout the empire for that purpose. Conquered tribes were made to accept the Inca as a descendent of the Sun and to join in prescribed worship. The year was organized for annual religious and agricultural festivals. Solar pillars near Cusco described by Spanish chroniclers were likely built to signal such events (Zuidema 1981b; Aveni 1981a).

6.3 Cosmology and Origins

Lake Titicaca lies on the border between Peru and Bolivia at an altitude of 3812 masl. It covers more than 58,000 square kilometers and is host to many islands, two of which figure centrally in Inca cosmology and origins (Figs. 6.4 and 2.11).

The Inca believed the Sun and Moon to have been born from Lake Titicaca and its *Isla del Sol* and *Isla del Luna*, respectively. The Sun was thought to have risen from an outcrop of rocks on the island that now bears its name and the Moon likewise from a smaller island in the nearby vicinity. The original Inca and his queen were royal descendants of the Sun and Moon and these islands became great huacas and state pilgrimage centers. Offerings were made amidst intricate ceremonies as Inca culture and religion grew around these origin myths (Bauer and Dearborn 1995).

The Sun of the Incas is further linked with water as their cosmology tells of its daily rebirth from the Vilcanota River in Peru (Fig. 6.5). After the western sunset the Sun travels beneath the Vilcanota and is therefore rejuvenated by the powers of its waters before being born again in the morning on the eastern horizon. During

Fig. 6.4 Lake Titicaca from Bolivia (Anthony Lacoste / CC BY / https://creativecommons.org/licenses/by/3.0)

Fig. 6.5 The Vilcanota River in October. (Author's photograph)

summer rains the river runs deep and fully charges the hot Sun. In the dry winters the Sun is dimmer and temperatures colder as the lower river waters cannot fully replenish solar powers (Urton 1981a).

6.4 Festivals and Ceremonies

Integral with the role of the Sun in Inca society were several prominent festivals. Inti Raymi, the festival of the Sun at the time of the June solstice, was one of the most significant of the Incas' year. The Incas commenced ceremonies for 8 days attended by the royal mummies and complete with sacrifices and great chanting. Gratitude was expressed to the Sun for present and future harvests (Hemming and Ranney 1982).

The time of the December solstice was the other great pinnacle of Inca solar worship with the festival of Capac Raymi (Dearborn et al. 1987). The pillars of Cusco, described by the chroniclers, could well have been used to forewarn of and celebrate both of these annual solar events.

6.5 Inca Horizon Astronomy

As Sun worshippers and followers of solar events, the Incas became masters of horizon astronomy (Sect. 5.6), the tracking of celestial events as they occur on the horizon. The Incas practiced horizon astronomy through positional observations of the rising and setting Sun on days of ceremonial and agricultural significance. As the Earth orbits the Sun, the tilt of its axis causes the horizon positions of sunrises and sunsets to move accordingly north or south ever so slightly each day. The Incas identified these astronomical events on the horizons of Cusco and monitored them with pillars (Aveni 1981a; Cobo 1990 [1653]; Zuidema 1981b).

Alternate methods of establishing sunrise positions may have been practiced at Machu Picchu and the neighboring ceremonial center of Llactapata, where the irregularities of the horizon could have provided natural calendrical markers for events such as the June solstice. In addition, a 33-meter-long corridor at Llactapata establishes a 4.3° window along the horizon that frames both the rising positions of the June solstice Sun and that of the Pleiades. A similar sighting device may have been established at the Coricancha of Cusco.

In *The Huarochirí Manuscript*, Salomon and Urioste (1991) relate that the Incas employed professionals called *yancas* to make such astronomical observations. Yancas observed the course of the Sun and its alignment with "calibrated walls" from designated vantage points (Salomon and Urioste 1991). Yancas were trained for solar and lunar observations and also observed and evaluated gnomon shadows.

Evidence suggests that related passage of time was recorded on quipus (Bauer and Dearborn 1995).

Solstices

The Sun's southernmost declination on its apparent ecliptic path of travel is currently S23.439°, and when it reaches this point we observe what is called the December solstice. In the Southern Hemisphere this is also known as the summer solstice and throughout the months leading up to it there are small southward movements of the Sun on the horizon each day, except for about 2 days before and after the solstice when there is no observable motion as the Sun "stands still" before reversing course back to the north (Bauer and Dearborn 1995; Urton 1981a; Sect. 5.3). The Sun's northernmost declination, N23.439°, occurs at the June, or winter, solstice when similar motions are observed, only in the opposite direction. The Sun travels on a lower arc in June but crosses high in the sky at the time of the solstice in December. In the region around Cusco there is a difference on the horizon of approximately 50° between the positions of the solsticial sunrises.

The Incas staged the great festivals of Inti Raymi and Capac Raymi at times of these prominent annual June and December solar events. "The winter solstice in the Andes marked the time when the Sun was reborn and the world revitalized, and the summer solstice marked the time of coming to naturity" (Salazar and Salazar 2014, p. 30).

Equinoxes

Like with solstices, there are two solar equinoxes. The March and September equinoxes occur when the Sun crosses the celestial equator, from south to north in March and from north to south in September (Sect. 5.3). In the Southern Hemisphere the September and March equinoxes signal the beginnings of spring and fall respectively and on those particular dates the hours of daylight and darkness are equal.

Andean tradition linked the equinoxes with fertility, September in the spring when the soil was first prepared for planting and in March, the fall, when maize was harvested. Bauer and Dearborn (1995) state that the festival of the Sun called Citua Raymi was celebrated every September at the equinox, and perhaps Pacha Pucuay Quilla at the equinox in March. The chronicles, however, do not offer significant support for Incan observation of equinox sunrises and sunsets (Zuidema 2007).

Zenith Sun

Between the times of the December and June solstices the Sun's apparent travel twice takes it directly over each geographical point lying between S23.44° and N23.44°. At the latitude of Cusco, about S13.5°, the two zenith passages occur on February 13th and October 30th (Zuidema 1981b, pp. 322–323; Sect. 5.5). On these dates the Sun passes directly overhead at local noon and vertical objects do not cast a shadow, a phenomenon of which the Incas were well aware. While this specific event took place overhead and not on the horizon, in their system of horizon astronomy the Incas included observations of the related positions of sunrise and sunset on days they knew to be of zenith passage (Zuidema 1981b). This recorded position might also have been used to determine the dates of nadir, or the anti-zenith Sun. Vertical towers, such as the Sunturhuasi in Cusco, were used to facilitate the observation of the Sun or Moon at zenith.

Chavín de Huántar, the Chavín site of the Late Initial, has been found to exhibit an alignment that coincides with the zenith sunrise and anti-zenith sunset. Burger (1992, p. 132) describes Gary Urton measuring the construction of the Old Temple and finding it to be oriented to 103.5° and 283.5°. The latitude of the site is at S09.6°, resulting in a flat horizon zenith sunrise of 099.6° and anti-zenith sunset of 279.6°. Burger states that when Chavín's horizon is taken into consideration the anti-zenith sunset would occur at 280.6°. Urton also observed that the setting of the Pleides at the time of construction would have been at 283.5°, the same alignment as the walls (283° 31′ walls; 283° 41' Pleiades).

Anti-Zenith Sun

Zuidema explored the concept of Inca observation of the anti-zenith passage of the Sun, the date that the Sun passes through nadir. The Sun cannot be seen at nadir, but Zuidema relates that the Incas observed solar horizon positions on these dates as well as those of the zenith passage (Zuidema 1981b; Sect. 5.5).

Zuidema states that dates of anti-zenith could be determined by reversing the direction of sunrise on days of the zenith Sun to arrive at the position of anti-zenith sunset. On days of anti-zenith in Cusco at S13.5° the Sun is at zenith at N13.5° (Zuidema 1989c). Zuidema maintains that a pillar of Cusco and several ceques were aligned to designate days of the anti-zenith Sun. Zuidema proposed a zenith/anti-zenith alignment between sucancas on Cerro Picchu and at Tipon, 24 km to the southeast. He states that the zenith sunrise could be observed from Picchu to Tipon and the anti-zenith sunset from Tipon to Picchu (Zuidema 1977; 1981b; 1982a; Ziółkowski and Kościuk 2018).

Zuidema (1982a) stated that the Incas were very interested in the dates of August 4th and August 18th. On August 18th the Sun goes through nadir, or anti-zenith passage. While the Sun cannot be seen at nadir, the full Moon closest to the 18th will pass at or very near zenith. The date of August 4th is nearly midway between the June solstice and the September equinox, specifically on August 5th. When a full Moon occurs on the June solstice another one will take place on August 18th, two lunations later.

Anti-zenith passage occurs in Cusco each August 18th and April 26th and coincides with the planting and harvest of maize, times of Inca ceremony and celebration (Zuidema 1981b; 1982a). Recording of zenith sunrises facilitated the calculation of anti-zenith dates. Maize-related agricultural festivals would likely be associated with anti-zenith observances.

Two structures were central to solar observations. The first, known as the ushnu, was a platform in the plaza of Huaycapata in Hanan Cusco that served as an observation point for the position of the Sun between two horizon pillars called Sucanca. Zuidema (1981b, pp. 320–321) states that the ushnu served to define the months of September and August. He adds that the Incas were interested in August and April as the respective beginning and end of their agricultural season and maintains that they could only have determined the corresponding dates of August 18th and April 26th through anti-zenith observance (Zuidema 1981b, p. 322).

A structure located on the central plaza called the Sunturhuasi figured centrally in zenith and anti-zenith observations (Fig. 6.6). It was located close to the ushnu and both are thought to have been aligned with the central of the four pillars of Sucanca on Cerro Picchu. The Sunturhuasi was a cylindrical tower with windows and a mast on its top that could have been used vertically to identify the zenith Sun and horizontally to observe the associated point of sunrise. The round tower of the Sunturhuasi helped to facilitate the prediction of the position of the anti-zenith sunset by reversing the zenith sunrise direction. Upon reaching the appropriate day the confirming anti-zenith observation could also be made from the Sunturhuasi. A second ushnu in Hurin Cusco may have been used in conjunction with the one in Hanan Cusco to establish the axis of zenith and anti-zenith. The Sunturhuasi was situated about 3° south of this axis. Zuidema asserts that the Sunturhuasi, the sucancas and the ushnus all likely had other purposes besides those that were astronomical (Zuidema, 1981b, pp. 323–324).

Fig. 6.6 Observation from the Sunturhuasi. (Guaman Poma, f. 329)

Bauer and Dearborn (1995, pp. 94–98) maintain that they find insufficient evidence to support anti-zenith celebrations by the Inca. They state that no direct historical data or archaeological remains have been found to support any Inca interest in the anti-zenith. Zuidema's (1981b; 1982a; 1989c; 2005; 2008a, 2008b) research and arguments are significant.

Vertical observations of the zenith Sun and horizontal observations of solar pillars for anti-zenith events appear to distinguish astronomical observational philosophy within the city of Cusco from the more solsticial orientations found with many of the outlying huacas of this study. Aveni (1981c) points out the brilliance of using the zenith-nadir axis and the horizon to determine the day of anti-zenith.

6.6 Architectural Alignments

Architectural structures sometimes played a role in Inca cosmology and were frequently designed with astronomical orientations. Certain structures were situated to use local geographic features with regard to solar alignments while others relied solely upon specific architectural designs. It is likely that in each such case those alignments with the Sun were carefully considered before construction began. There is also evidence that alignment with the rise of the Pleiades was observed and that such an orientation served to guide the viewer's eyes to the desired celestial object. Zuidema and Aveni examined the Coricancha of Cusco and found it to be oriented in this manner (Figs. 6.3 and 6.7). Their measurements indicated an alignment for the Pleiades that passed between two western rooms, across a ceremonial basin, between holes for gold and precious stones, and then through a space between two eastern rooms. The Coricancha is thought to have been a walled enclosure of stone structures with thatched roofs (Salazar and Salazar 2014). Zuidema and Aveni found the alignment of the Pleiades with the western and eastern walls of the Coricancha to be to an azimuth of 66° 44′. Their calculations further show that the Pleiades rose at 65° 58′ in AD 1500; 66° 22′ in AD 1400; and 66° 46′ in AD 1300 (Zuidema 1982a pp. 212–214. In later research Ziólkowski and Kościuk (2018) find little to support precise astronomical observation at the Coricancha. They do support precise observations in one small window, but dispute much of the other work.

Many structures throughout the Incas' former empire display evidence of celestial orientations. Certainly, the many temples of the Sun received such considerations when they were built. Two additional examples are the Torreon of Machu Picchu and the Sun Temple at Llactapata, which each have orientations toward the rise of both the June solstice Sun and of the Pleiades.

The sophisticated astronomy of the Incas was dependent upon precise observations of events on the horizon. In this effort they employed windows aligned for the Sun and solar pillars. While the towers on the horizons of Cusco no longer remain, Sect. 8.7 includes study of two such pillars above the modern village of Urubamba.

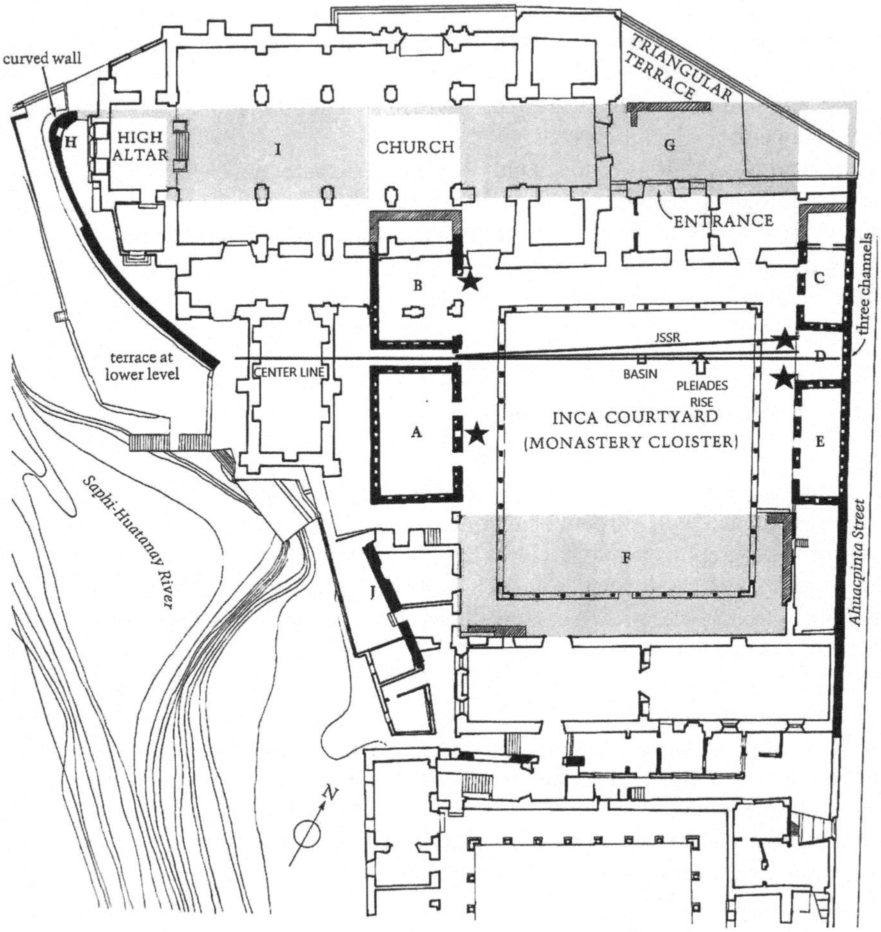

Fig. 6.7 Plan of the Coricancha. The sightline identified by Zuidema for the Pleiades rise passes between two western rooms, across a ceremonial basin and between two eastern rooms. The encircled stars indicate positions to be used for gold or precious stones. (From Zuidema, 1982a, p. 213)

6.7 Ushnus, Sucancas, Pillars, and Gnomons

Zuidema (1989c, pp. 406–407) states that observations of Cusco's horizon were made with three types of constructions: two ushnus, one near the Sunturhuasi and the other further to the southeast; the Sunturhuasi, a building that served as a state gnomon; and the gnomon, a vertical pole used mainly to measure the Sun on the day of zenith passage. He asserts that precise antizenith sunset observations were made with pillars on Mount Yahuira (Cerro Picchu) from the ushnu at the Plaza of Hanan Haucaypata (Zuidema 2007; Fig. 6.8). The ushnus joined the Sunturhuasi for observations of the central horizon, approximately east and west (Zuidema 1989c, p. 404).

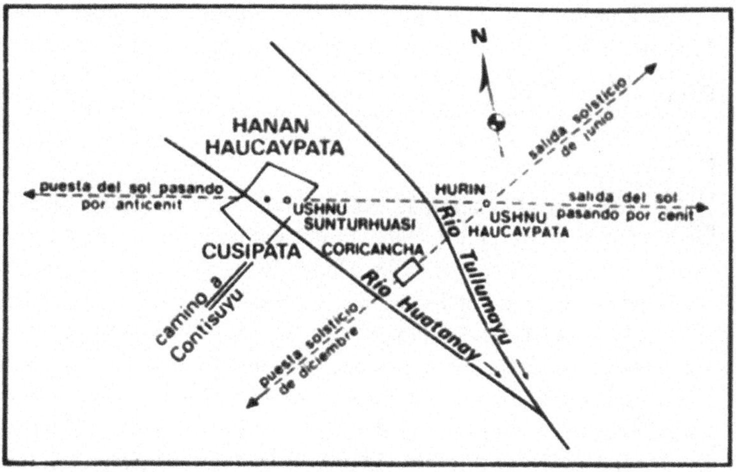

Fig. 6.8 In Cusco the Coricancha and the Hurin Huacaypata Ushnu were aligned for the axis of the June solstice sunrise and December solstice sunset. The Hanan Haucaypata Ushnu, the Sunturhuasi, and the Hurin Huacaypata Ushnu were aligned for the axis of the zenith sunrise and the antizenith sunset. (From Zuidema 1989c, p. 407)

Ushnus

Ushnu is a Quechua word meaning the concept of a specific space created for the use of the emperor or certain other government officials. The status assigned to such a platform allowed for visual affirmation of divine and state authority. Ushnus were built away from Cusco as part of the effort to exercise control over the populations of conquered territories. "[Pachacuti] also directed that there be a throne and seat of the Incas called *usno* in each *uamani* [district]" (Guaman Poma 2009 [1615], p. 205). Morris and Thompson (1985, p. 59) discuss the Ushnu located at the site of Huánaco Pampa and state that "Large ushnu platforms may still be observed at Vilcas Waman, in the Peruvian south-central highlands, and at Pumpu, on the shores of Lake Junin. They are all roughly pyramidal in shape and associated with open plaza areas, although none are exactly alike or closely similar in detail." They describe the ushnu at Huánaco Pampa as a platform sitting on top of two lower platforms with steps on its side to the south. Another potential ushnu exists at Saihuite and is shown in Fig. 7.80.

Ushnus were platforms for making offerings, for observing visual sightlines, and for receiving liquid offerings, such as chicha, made to the Pachamama. The Inca sat atop the ushnu when he dispensed justice (Zuidema and Quispe 1973). Some ushnus were not huacas – they were not prayed to or asked for advice and guidance. "The Incas have places called *usnos* designated throughout this kingdom for sacrifice, which is always for the *capacocha* to the Sun and the *huacas* .... The Inca made sacrifices to his father the Sun with gold, silver, handsome ten-year old boys and girls who had no blemishes, not even a mole" (Guaman Poma 2009 [1615], p. 201).

An ushnu was located on the Plaza of Huaynapata in Cusco. Zuidema (1989c, p. 453) states "The *ushnu* was the architectural center of the plaza of Cuzco and joined with the Temple of the Sun influenced the outline of the general plan of the city." "The Ushnu itself is a very elusive concept; its relation to verticality in the Inca cosmos has been fully developed by Zuidema [1977-8], who has demonstrated that it was manifested in a multitude of material objects including an altar of sacrifice, a platform of stones, a mojon, and even a hole in the earth" (Aveni 1981a, p. 313; Figs. 6.9 and 6.10).

Huacas were animate or inanimate objects that were specially venerated. As described in Sect. 4.3, they could be many things including such as natural features of the landscape, trees, springs, and rocks. Huacas were thought to be possessed by the local deity whose essence of force gave prosperity to those around it (Staller 2008, pp. 272–274) and received commensurate worship and offerings.

While an ushnu imbued the divine power of the empire to those seated upon it, it differed from a huaca in that it did not emanate prosperity and was not worshipped. An ushnu was a platform with several tiers and a staircase on one of its sides. Its purpose was for the king or other prominent government officials to view or perform rituals. Ushnus figured prominently in ceremonies regarding the sanctification of divine rule (Staller 2008, p. 285), and it also played a role in Capac Hucha, or child sacrifice (Guaman Poma 2009 [1615], p. 201).

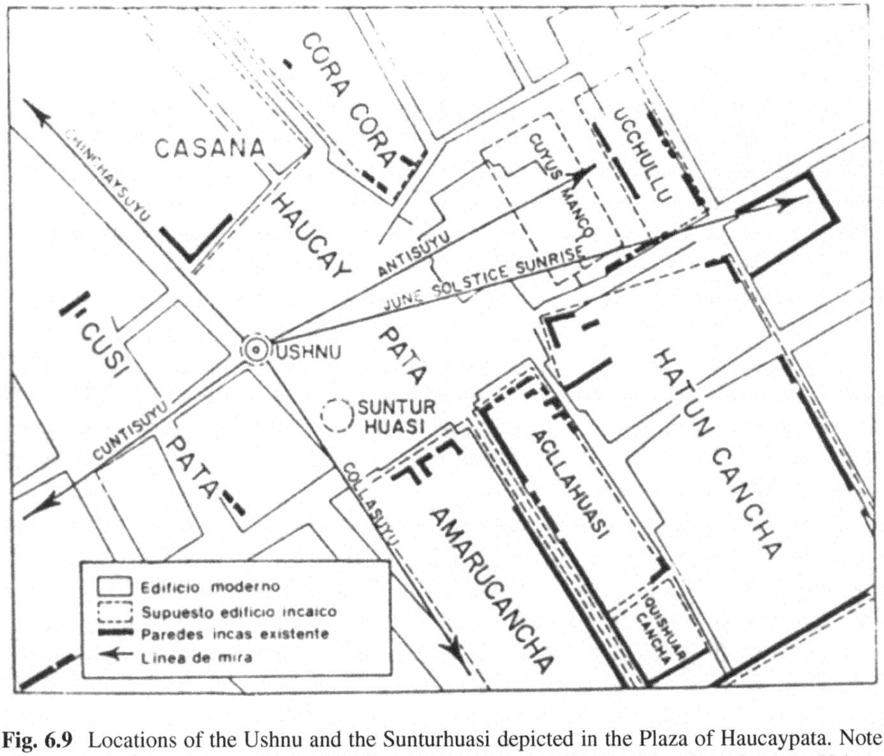

Fig. 6.9 Locations of the Ushnu and the Sunturhuasi depicted in the Plaza of Haucaypata. Note the axis of the June solstice sunrise and that the four suyus of Cusco each radiate from the Ushnu. (From Zuidema 1989c, p. 431)

Fig. 6.10 Drawing of the Ushnu. (Guaman Poma, f. 160)

Zuidema (1989c) found the three main elements common to ushnus to be that they were multi-tiered platforms, they were associated with pillars, and they had basins where chicha or water could be deposited.

Sucancas

Zuidema (2005) relates that according to Polo de Ondegardo the pillars of Pucuy Sucanca marked the start of the rainy season while the pillars of Chirao Sucanca conversely indicated the beginning of the dry part of the year. He maintains that other pillars were erected to establish the time of various festivals and even indicate the passage of months. Pucuy Sucanca and Chirao Sucanca were each huacas as a part of the Cusco ceque system (Zuidema 2005).

Zuidema (1981b) gives an excellent discussion of these two sucancas, which appears to be a much more satisfactory explanation than that suggested by Bauer and Dearborn. He includes both sucancas and sayhuas in his description. *Sucancas*, by definition, were sunrise or sunset points and the pillars of Pucuy and Chirao helped to define the axis of the zenith sunrise and the anti-zenith sunset. *Sayhuas* were markers and these near Cusco also used the position of the Sun on the horizon, in this case to denote the start of Incaic months. Zuidema points out that this system likely included 12 sayhuas and two sucancas. The Chirao sucanca was likely located on Cerro Picchu to the northwest of Cusco and the Pucuy sucanca far to the south-east on a mountain in *Quispicancha*, or Tipon. Zuidema continues that the names Pucuy and Chirao referred to seasons – *Pucuy* to "the time when fruits ripen during the latter part of the rainy season" and *Chirao* to "the time when the first changes in the weather occur after the dry season" (1981b, p. 330). Solar alignment with the Pucuy sucanca therefore likely indicated a date in February, also the time of the zenith sunrise, and the Chirao sucanca in August, corresponding with the anti-zenith sunset.

Zuidema (2008b, pp. 252–254) describes four sucancas, three identified as such in the Cusco ceque system plus Quiancalla, whose usage appears very similar to the other three. Two of these are those mentioned previously as Pucuy and Chirao. Two sucancas were east of the city, the first near Cerro Mutu and the other, Chirao, near Cerro Quispicancha. The third, Pucuy, was to the west of Cusco on Cerro Yuhira, now known as Cerro Picchu. Quiancalla, the last, was also on the west side of the city. The relatively close western sucancas had pillars, while the more distant eastern ones did not.

Quiancalla was observed at the time of the June solstice sunset from north of Cusco at Chuquirmarca on a hill called Maturcalla. The sucanca near Cerro Mutu served to observe the December solstice sunrise from the temple of Puquincancha to the south of Cusco (Zuidema 2008b, p. 255). The zenith sunrise and anti-zenith sunset were observed between the Chirao Sucanca on Cerro Picchu and the Pucuy Sucanca at Cerro Quispicancha. The sucanca on Cerro Picchu also was observed from the ushnu of the Hanan Huaycapata plaza at the time of the anti-zenith sunset. The sucancas all played a role in ritual movements during the month of the June solstice, in the second month before the December solstice, and in the second month following the December solstice (Zuidema 2008b, pp. 256–260) (Fig. 6.11).

Pillars

The Incas were devoted students of astronomy and likely tracked sunrises and sunsets, the Moon, and certain stars whenever possible with existing geographical features. Some horizons presented a challenge to precision, however, that apparently was solved with pillars built to mark annual positions. Cobo (1983 [1653], pp. 251–252) related that pillars were erected on the horizon of Cusco to track solar movement. These structures no longer remain, but two such Sun pillars are located at 3860 meters on a ridge above Q'espiwanka (Fig. 8.27), the palace of Huayna Capac, and are aligned to mark the rising of the June solstice Sun when viewed from a large boulder located in the center of the palace courtyard. There is ethno-historical description by Spanish chroniclers of pillars surrounding the city of Cusco, but none survived the Spanish purge of idolatry. These towers, near the modern community of Urubamba, provide evidence of this type of Inca celestial alignment,

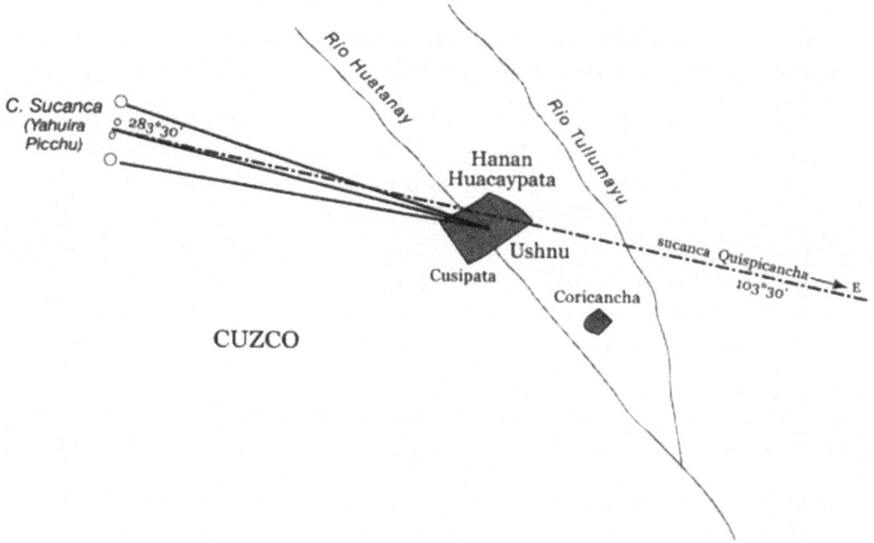

Fig. 6.11 The Chirau Sucanca on Cerro Picchu was used in the observation of the anti-zenith sunset from the Ushnu at the plaza of Hanan Huacaypata; it was also used in conjunction with the Pucuy Sucanca at Cerro Quispicancha in the observation of the zenith sunrise and the anti-zenith sunset. (From Zuidema, 2008b, p. 250)

demonstrate that such solar pillars did exist for the purpose of marking significant Inca astronomical events, and add credibility to the colonial reports of similar structures on the horizons of Cusco. Other examples have been identified above the Isla del Sol (Dearborn et al. 1998) and near Puncuyoc (Bernard Bell, personal communication, 2010).

There are differing accounts as to how many pillars actually surrounded Cusco. Betanzos (1996 [1576], p. 68) describes eight pillars in two sets of four – one set where the Sun comes up and the other where the Sun goes down. Cobo (1983 [1653], pp. 251–252) states that there were four pillars – two on the eastern and two on the western sides of Cusco. He continues that there were additional markers placed on the horizon where the Sun would reach each month. Garcilaso (1961 [1609], p. 71) says that there were "eight towers on either side of the city of Cuzco, four of which faced the rising, and four the setting sun." His wording can be misleading, but he describes a combined total of eight towers on the two horizons.

Aveni, with Zuidema, identified pillars and observing axes from three locations in Cusco:

(a) A pair of pillars which mark the June solstice sunset point as viewed from Lacco, a complex of rock carvings on a hill north of Cuzco.
(b) A pair of pillars to mark the December solstices as seen from Coricancha, the center of the ceque system.

(c) The sucanca, four pillars situated on Cerro Picchu to mark time for the planting season, centered on the place where the sun sets on the day of passage through the antizenith (Aveni 1980, p. 303)

The pillars were viewed from the ushnu in the plaza of Hanan Huacaypata. The zenith/antizenith axis also passed through the ushnu in the plaza of Hurin Huacaypata (Fig. 6.8). The Sunturhuasi was close to this axis and also close to the Hanan Huacaypata ushnu (Aveni 1981a). The Incas would not necessarily have had to understand the physical concept of nadir passage of the Sun to have made and utilized such an observation (Zuidema 2008b) (Fig. 6.12).

Cobo (1990 [1653], pp. 59, 60–61, 83) supports this with his descriptions of corresponding huacas (Fig. 6.8):

> [Ch 6:9] The ninth *guaca* was a hill named Quiangalla, which is on the Yucay Road. On it were two markers or pillars which they thought denoted the beginning of summer when the sun reached there.

> [Ch 8:7] The seventh was called Sucanca. It was a hill by way of which the water channel from Chinchero comes. On it there were two markers which indicated that when the sun arrived there they had to begin to plant the maize. The sacrifice which was made there was directed to the Sun, asking him to arrive there at the time which would be appropriate for planting, and they sacrificed to him sheep, clothing, and miniature lambs of gold and silver.

> [Cu 13:3] The third, Chinchincalla, is a large hill where there were two markers; when the sun reached them, it was time to plant.

Zuidema suggests that Lacco is the most likely candidate for *Chuquimarca* (Aveni 1981a). Cobo (1990 [1653], p. 65) gave the following description of this huaca:

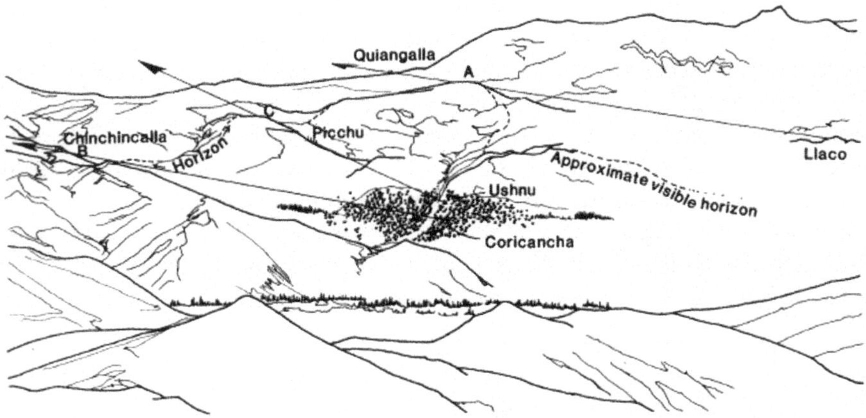

Fig. 6.12 (**a**) The pillars of Quiangalla viewed from Lacco, (**b**) the pillars of Chinchincalla viewed from the Coricancha, and (**c**) the sucanca of Cerro Picchu viewed from the ushnu of Hanan Haucaypata. (From Aveni 1981c)

> [An 3:4] The fourth was called Chuquimarca; it was a temple of the Sun on the hill of Mantocalla, in which they said that the Sun descended many times to sleep. For this reason, in addition to everything else, they offered it children.

An anonymous chronicler c. 1570 describes the ushnu:

> And when the sun stood fitting in the middle between the two pillars they had another pillar in the middle of the plaza, a pillar of well worked stone about one estado high, called the Ushnu, from which they viewed it. This was the time to plant in the valley of Cuzco and surroundings. (Aveni 1981a, p. 311)

Soon after the conquest Spanish priests began a campaign to eradicate whatever supported the indigenous religion of solar worship, including the means to track movement of the Sun on the horizon (Garcilaso 1961 [1609], p. 73). Ultimately all of the solar pillars of Cusco were destroyed, but the aforementioned pair above the modern community of Urubamba survived. These are discussed more fully in Sect. 8.7 (Figs. 6.13 and 6.14).

Fig. 6.13 Drawing depicting pillar construction. (Guaman Poma, f. 352)

Fig. 6.14 Drawing depicting Cusco pillars. (Guaman Poma, f. 354)

Gnomons

The Incas used gnomons for shadow effects, solar tracking, and related calculations. Horizontal gnomons, such as those found at the Incamisana in Ollantaytambo (Figs. 8.60 and 8.61), were carved to cast images in shadow at certain times of the year. Vertical gnomons were used to judge time or direction, such as indicating zenith passage of the Sun by the withdrawal and disappearance of the gnomon's shadow.

Pillars functioned as gnomons (Staller 2008, p. 287), were viewable by all, and could have been intended for public spectacle as well as solar measurements. The observance of shadows cast by gnomons, however, was not practical for large groups and therefore their use was more probably confined to small gatherings of elites making solar determinations (Bauer and Dearborn 1995).

6.8 Inca Calendar

Time reckoning by lunar months was practiced in the Andes as it was a simple exercise for anyone to count and keep track of lunar cycles. Garcilaso de la Vega (1961 [1609], p. 71) relates that the Incas counted months by Moons. The Incas used a synodic lunar calendar measured by Moon phases and rituals falling between solar observations were determined by the Moon (Zuidema 2007). Events associated with the Sun were tied to the lunar calendar as such celebrations required the observance of a new or full Moon in temporal proximity to the solar occurrence (Zuidema 1981c). Pillars tracked the months of the Sun-based calendar while lunar phases were incorporated for festivals by proclaiming the first new Moon of the solar month to have the same name as that month (Bauer and Dearborn 1995; Zuidema 2007). The year was divided into "two equal periods, four nearly equal periods, twelve periods of rather unequal length, and 41 very unequal periods" (Zuidema 1982b, p. 61). To define the calendar the Incas use six rising or setting horizon positions:

1. Sunrise on May 26
2. Sunrise at zenith passage
3. Sunrise at December solstice
4. Sunset at December solstice
5. Sunset at anti-zenith passage
6. Sunset at June solstice (Zuidema 1982b, p. 73)

"Moons," or lunar cycles, have long been used by ancient peoples to define months as they follow a readily recognizable pattern. The Incas were no exception and the Quechua language employs only one word, *quilla*, for both month and Moon. Bauer and Dearborn (1995, pp. 30–31) relate the following names of month equivalents according to the chronicler Guaman Poma de Ayala:

January	*Capac Raymi*
February	*Paucar Uaray*
March	*Pacha Pocay Quilla*
April	*Ynca Raymi*
May	*Atun Cusqui*
June	*Cuzqui Quilla*
July	*Chacra Conocuy*
August	*Chacra Yapuy Quilla*
September	*Coya Raymi Quilla*
October	*Uma Raymi Quilla*
November	*Aya Marcay Quilla*
December	*Capac Ynti Raymi*

Zuidema (1977) discusses a method of time tracking in Cusco related to Inca huacas and ceques (Sect. 4.6). He finds in the chronicles that the ceque system was a means for counting days of the year with each of the 328 huacas used in order to mark daily passage of time. Ceques were grouped by threes and each of these groups represented 1 month (Zuidema 2007). He suggests the remaining 37 days in a 365-day tropical solar year equate to the approximate period the Pleiades is invisible by proximity to the Sun between 3/4 May and 8/9 June. He continues that the 41 ceques could be used in association with the passage of 41 eight-day weeks.

When seeking to compare dates in the Inca Empire with those of the Gregorian calendar it must be remembered that when the Vatican first instituted the Gregorian calendar in 1582 (10 years after the execution of Tupac Amaru at the end of the Spanish conquest) it necessitated an adjustment that deleted 10 days in October of that year. The same adjustment was ordered in Peru in October 1584. Historical accounts before this that have been given Gregorian dates must add 10 days to be accurate.

6.9 The Moon

Lunar cycles of approximately 29.5 days divided the year in the Andes. The solar calendar instituted by the Incas retained similar monthly subdivisions and likely was periodically intercalated.

Inca cosmology describes the Sun as first rising from the Island of the Sun in Lake Titicaca. It also relates the Moon's first ascent from the nearby Island of the Moon. Both islands were important state pilgrimage and worship centers (Bauer and Stanish 2001). The ruling Inca was the son of the Sun and his wife the daughter of the Moon. The Sun was male and the Moon female, the female association made perhaps due to the Moon's relation with the female menstrual cycle that falls between the sidereal and synodic lunar cycles (Urton 1981a).

Bauer and Stanish (2001) also tell us that the Spanish chronicler Polo de Ondegardo recorded that offerings were given to the Moon both at the time of eclipses and during childbirth and it was the woman's role to do so. They continue that Garcilaso recorded that a hall of the Coricancha was dedicated to the Moon and the Moon's image was placed in it along with mummies of the empire's dead queens. The Moon was called *Pacsamama* (Moon Mother) or *Mamaquilla* (Mother Moon).

Urton (1981a) found dark spots seen on the surface of the Moon to have names and significance. Such were as a woman and a llama or a woman holding her baby daughter. Urton compares these images as similar with those the Incas placed in the darkened parts of the Milky Way.

Certain crops were planted with regard to lunar cycles. Maize was sown with a waxing Moon, from new to full, but potatoes were to be planted while the Moon was waning. The lighted portion of the Moon was called *pura* and the dark portion *wañu*. In general terms underground crops were sown while the Moon was waning, from pura to wañu, and aboveground crops while waxing from wañu to pura (Urton, 1981a).

Zuidema (1981b) observed that the full Moon is at zenith when the Sun is at nadir and has found this to be related to the Inca's timing of the anti-zenith Sun.

6.10 The Milky Way

Inca cosmology viewed the Milky Way as a river flowing across the night sky in a very literal sense. Incas saw earthly waters as being drawn into the heavens and then later returned to Earth after a celestial rejuvenation. The Earth was thought to float in a cosmic ocean and when the "celestial river's" orientation was such that it dipped into that ocean the waters were drawn into the sky. "The Milky Way is therefore an integral part of the continuing recycling of water throughout the Quechua universe" (Urton 1981a, p. 60).

Orientation and Quadripartition

The Milky Way passes brightly overhead at southern latitudes such as that of Cusco and the Incas observed it closely (Urton 1981a). They saw it as two separate rivers, due to the Earth's rotation and the Milky Way's alternating position on the horizon each 12 hours. The plane of the Milky Way is inclined between 26° and 30° with the axis of the Earth's rotation. This orientation is 26° degrees toward the south celestial pole and 30° toward the north (Urton 1981a). The Milky Way at times will be viewed as rising in the southeast, passing through the zenith, and setting in the northwest. Twelve hours later the horizon positions have shifted and the band of stars rises instead from the northeast, traveling again through the zenith, but now setting in the southwest. This 24-hour rotation cycle creates two zenith-intersecting intercardinal axes that divide the celestial sphere into four observable quarters (Urton 1981a; Fig. 6.2).

The Milky Way risings are interesting because of correlations with the intercardinal axes and the four points of solstice horizon events. At the time of the December solstice, when the Sun rises at 114° on the Cusco horizon, the evening positioning of the band of the Milky Way lies similarly to the southeast. During the June solstice sunrise at 064° the Milky Way is situated in like fashion in the northeast. Times of the solstices are the only ones when the Sun rises and travels with the Milky Way (Urton 1981a). Both the celestial river and the Sun rise together with the dry season in June and the rainy season in December (Urton 1981a) and the villagers in Urton's study use this correlation to explain the seasonal intensity of the Sun, which feeds upon the powerful waters.

The Incas may have ordered their sky by this celestial quadripartition, in contrast to an orientation more related to the general direction of ecliptic travel that was used by other ancients. There is no definitive evidence to confirm this, however. Urton (1981a) said that this perspective gave those in the Andes a nearly 90° difference in their orientation to the heavens, and cosmological constructs were developed accordingly. He asserts that the primary axis for celestial references was east-west, rather than the north-south one common in systems of the Northern Hemisphere. The quadripartition also appears to have influenced orientations on Earth. Urton found evidence that the two primary trails of Misminay follow the same terrestrial axes as those above in the night sky.

Celestial River

Andean cosmology ties the Milky Way with the Vilcanota River. The Vilcanota flows southeast to northwest through the Sacred Valley, past Machu Picchu and beyond. Its waters are thought to rise into the Milky Way and, once having traveled its celestial course, fall again to the Earth as rain. The Sun is stronger in the summer because it drinks from the swollen Vilcanota as it travels beneath it at night. It is weaker in the winter because it has had less to drink (Urton 1981a, p. 69). The Milky Way is said to be a heavenly reflection of the Vilcanota.

Dark Constellations

The Milky Way provided visual inspiration for several themes of Inca cosmology. The Incas recognized *dark constellations*, or the shapes of beings formed by dark *clouds* in the visible band of the galaxy. These dark spots in the glow of the Milky Way's stars are formed by interstellar dust blocking the light from the bright array of stars behind it. The Incas saw great cosmological characters meant to guide them in their daily lives.

The dark constellations are most visible at the time of the March equinox when they span the band of the Milky Way at midnight. During the September equinox the fewest will be seen as they have since rotated to be beneath the horizon for much of the night. At the solstices the Milky Way briefly aligns with the horizon points of the respective solsticial sunrise and sunset and the animals of the dark constellations appear to follow the solar path on these significant occasions (Urton 1981a). Urton finds the dark constellations to be related to the rainy season and asserts that "it is ... essential to study the Dark Cloud constellations by analyzing the connections between sky and water and earth and water" (Urton 1981a, p. 173) (Fig. 6.15).

Fig. 6.15 Inca Dark Constellations of the Milky Way – the procession from right to left: (1) Machacuay the serpent, (2) Hanp'atu the toad, (3) Yutu the tinamou, (4) Yacana the mother llama, (5) Uñallamacha the baby llama, (6) Atoq the fox, and (7) Yutu the second tinamou. (From Milky Way watercolor by Jessica Gullberg with Urton's constellations)

Following are descriptions of these primary dark constellations, from right to left, as viewed during my field research.

Machacuay the Serpent (1)
Serpents figure prominently in Inca cosmology and are the creatures representing Uchu Pacha, the underworld and lowest of the three worlds of Inca existence. Machacuay leads this dark celestial procession as the constellations move left to right across the night sky. Van de Guchte (1990) says that amaru, or serpents, emerged from their underworld environs via rivers and are thought to be related to rainbows and foretell of rain. Machacuay can be seen at the beginning of the rainy season. The serpent's dark figure is long like a snake and travels head before the tail (Urton 1981a).

Hanp'atu the Toad (2)
Hanp'atu follows closely behind Machacuay. Toads were thought of as bad omens as they were created by the devil. Hanp'atu is a much smaller dark section of the Milky Way to the left of the snake (Urton 1981a).

Yutu the Tinamou (3)
Tinamou are birds indigenous to the Andes and are of very ancient lineage. Yutu follows Hanp'atu in the Milky Way and likewise is much smaller than Machacuay. This Yutu is adjacent to the constellation of the Southern Cross, is at zenith on the December solstice, and nadir on the June solstice. Yutu is known as the *Coalsack* in Western astronomy (Urton 1981a).

Yacana the Llama (4)
Llamas figure prominently in many aspects of Inca culture and this celestial figure was thought to animate the llamas on the Earth (Salomon and Urioste 1991). Yacana is a constellation much larger than Hanp'atu or Yutu and dominates the Incas' dark constellation section of the Milky Way. Yacana is situated between Centaurus and Scorpius. The prominent stars α and β Centauri serve as the llama's eyes and as such are known as *Llamacñawin*, the "eyes of the llama" (Urton 1981a; Fig. 6.16).

Uñallamacha the Baby Llama (5)
Below Yacana is a smaller dark constellation called Uñallamacha that is said to be a cria, a baby llama, suckling its mother (Urton 1981a; Fig. 6.16).

Atoq the Fox (6)
Following Yacana and Uñallamacha in the sky from the left is the somewhat smaller constellation of Atoq, the fox. Atoq lies on the ecliptic between the constellations of Scorpius and Sagittarius and the Sun enters it during the December solstice. Urton (1981a) relates that the Milky Way and Atoq catch up and rise with the summer solstice Sun in the southeast during the same period of time that terrestrial baby foxes typically are born. "... the sun rises into [Atoq] ... from about December 15 to December 23" (Urton 1981a, p. 70).

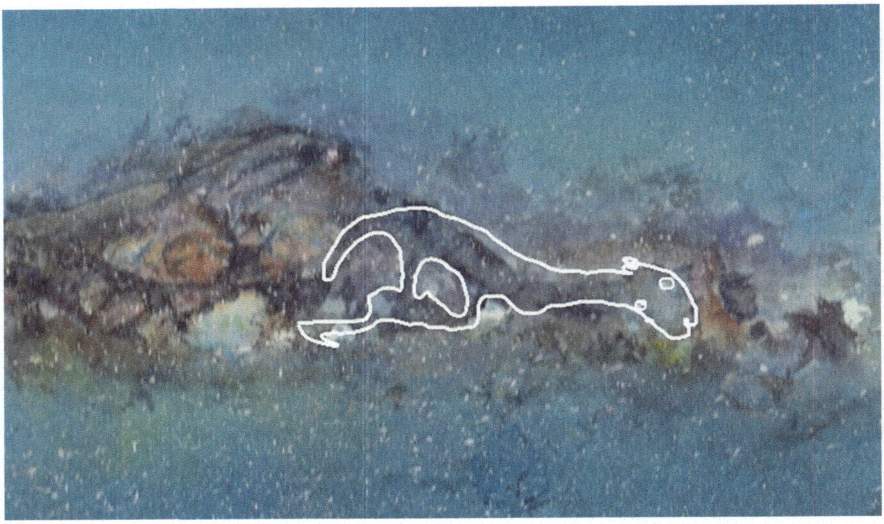

Fig. 6.16 Enlarged example of the mother (4) and baby (5) llamas; the mother's head is near the Coalsack. (From Milky Way watercolor by Jessica Gullberg with Urton's constellations)

Yutu the Tinamou (7)
Urton (1981a) lists a second constellation called Yutu. This additional tinamou follows Atoq and completes the celestial procession.

Shepherd
While Urton describes a second Yutu after Atoq, a painting by Miguel Araoz Cartagena of the dark constellations displayed within the Coricancha emphasizes a shepherd, Michij runa, following the fox instead.

Puma
Giulio Magli (2005) with reference to Urton (1981a) suggested as well a puma dark constellation with a location following Yutu between Cygnus and Vulpecula. He continued that the city plan of Cusco might represent this celestial puma.

6.11 Stars

The people of the Andes, like other cultures around the world, gazed at the heavens in wonder and found imaginary shapes in the stars above. For the Incas, however, these constellations were only a part of a celestial landscape that was filled with beings found in many astral shapes and meteorological phenomena. Spanish accounts speak of stars within a zodiac, but this was from a European perspective not shared in South America. The Incas often did not distinguish between planets and the brightest stars. The Quechua word *chasca* was used for both, while *coyllur* was the term for stars of lesser prominence (Bauer and Dearborn 1995).

The Incas believed there to be a close association between animals on the Earth and in the sky and that celestial counterparts had influence over the health and reproduction of those below. Many names for stars existed as the Incas did not impose their astronomical taxonomy on the peoples they conquered. The Incas went to great lengths to define worship of the Sun, but left stellar mythology largely unchanged (Bauer and Dearborn 1995).

Planets

Planets were likely recognized among the stars, though observation of the planets is not well documented in Inca astronomy. Venus is noted, however, both as an evening and as a morning star. The planet is bright enough to be seen in the daytime, but never crosses the sky at night. Bauer and Dearborn (1995, p. 138) point out that Venus was referred to by many names such as *Chasca Cuyor, Pacaric Chasca, Pacari Cuylor, Auquila, Pachahuárac, Chachaquaras,* and *Atungara.*

Constellations

Constellations have been traced in the night sky by most every culture that has observed the heavens. Evidence has been found to support that certain star groupings, such as the Hyades, the Pleiades, or Orion have been recognized many times. Spaniards attempted to define Andean astronomy in European terms, failing to fully realize that the Incas viewed the cosmos in a different way.

In addition to the dark constellations previously discussed, the Incas also recognized certain star to star groupings. While European astronomy followed a zodiac that centered around the ecliptic, the Incas oriented their sky with the Milky Way. Their star constellations are located in close proximity, especially between Taurus and Orion where the Milky Way is less brilliant. While dark constellations primarily represented animals, star to star constellations were more often objects (Urton 1981a). Cobo (1990 [1653], pp. 30–31] writes of the Incas designating corresponding stars as patrons for each animal species. The stars thusly were worshipped by various groups with vested interests.

Examples of Andean constellations are as follows (Bauer and Dearborn 1995; Urton 1981a). Given are the name, an English translation, and the location as described by the associated Western constellation or stars:

Santissima Cruz
"The Sacred Cross"
α, β, δ, π, and σ Scorpii

Amaru
"The Serpent"
Scorpius

Llamacñawin
"The Eyes of the Llama"
α and β Centauri

Huchuy Cruz
"The Small Cross"

The Southern Cross

Hatun Cruz (1)
"The Great Cross"
Rigel, Sirius, Procyon, and Betelgeuse

Hatun Cruz (2)
"The Great Cross"
Procyon, Castor, η and μ Gemini

Collca (1)
"The Storehouse"
The Pleiades
(Also Qutu, The Pile)

Collca (2)
"The Storehouse"
The Hyades
(Also Pisqacollca, The Five Storehouses)

Chakana (1)
"The Bridge"
Orion's Belt

Chakana (2)
"The Bridge"
ε, δ, and η Canis Major

Pachapacariq Chaska (1)
"The Venus of the Western Suyu"
Canopus

Collca (3)
"The Storehouse"
η, θ, ι, κ, λ, and υ Scorpii

Pachapacariq Chaska (2)
"The Venus of the Northern Suyu"
Altair

Urcuchillay
"Llama of Many Colors"
Lyra

Catachillay
"Female Llama with Lamb"
Altair and Deneb

Chasca Cuyllor
"Morning Star" "Evening Star"
Venus

The Pleiades

The Pleiades were of great importance in Inca astronomy. The Incas worshipped the Pleiades: they appear to have viewed them as a prognosticator of life and death.

On a practical level, the Incas may have found them useful in predicting times for planting and harvests (Fig. 6.17). Cobo (1990 [1653], p. 30) states the Pleiades were called *Collca* and that the "power that conserved the animals and birds flowed from this group of stars." The Pleiades disappear behind the Sun for approximately 37 days and first return into view, their heliacal rise, around June 9th, about 12 days before the solstice sunrise. The date of the heliacal rise is somewhat a function of the Moon's phase and atmospheric conditions, and from six to nine stars might be visible (Bauer and Dearborn 1995).

The stars in this prominent grouping were viewed with regard to their relative brilliance. A bright predawn appearance of the cluster near the time of the June solstice indicated a future season of rain followed by an ample harvest. A dull appearance, due to atmospheric obscuration by high cirrus clouds, indicated there would be drought in the months to come. Potatoes have shallow roots and are sensitive to drought. When the dimmest of the stars of the Pleiades are visible, the apparent diameter of the asterism grows by approximately 25% (Orlove et al. 2000). The relative position of the Pleiades' brightest star, Alcyone, appears to shift according

Fig. 6.17 The Pleiades are a prominent asterism observed not only by the Incas, but other ancient cultures as well; this is their orientation as viewed in the Southern Hemisphere. (NASA, ESA, AURA/Caltech, Palomar Observatory / https://commons.wikimedia.org/wiki/File:Pleiades_large.jpg)

to the number of the other stars that can be seen – this also was used for forecasting. Alcyone appearing toward the front of a large and bright Pleiades indicated early planting in late July-September. When the Pleiades appeared smaller and duller, with Alcyone apparently shifting more toward the rear, this indicated that planting should be delayed until October-November, closer to the annual rainy season (Urton 1981a). In actuality, a method had been discovered for anticipating the arrival of El Niño in a manner still used today. Orlove, Chiang, and Cane (2000, p. 68) stated: "We find that poor visibility of the Pleiades in June – caused by an increase in sub-visual high cirrus clouds – is indicative of an El Niño year, which is usually linked to reduced rainfall during the growing season several months later."

Contemporary Peruvian and Bolivian farmers continue to use this traditional method of crop management. While their observations can be made on several dates, a prominent one is June 24th, the Day of San Juan (St. John's Day), well after the heliacal rise of the Pleiades. This method has likely been used for hundreds of years and similar observations could have been made by the Incas at sites with alignments, such as those at the Coricancha (Sect. 6.1), Machu Picchu's Torreon (Sect. 9.2), and the Llactapata Sun Temple (Sect. 9.4), but it has not been conclusively established that they performed this type of weather forecasting (Krupp 2008; Urton 1981a; Orlove et al. 2000; Malville et al. 2006).

Cobo (1990 [1653], p. 61) writes of a name associated with the Pleiades, *Catachillay*, in his huaca descriptions:

> [Ch 8:10] The tenth *guaca* was called Catachillay. It is a fountain which is in the first flat place that descends to the Road of Chinchaysuyu.

Zuidema (1982a, pp. 211–214) states the ceque with this huaca includes the pillars of Sucanca and can indicate the horizon position of the last heliacal set of the Pleiades on or about April 15th. He continues that this observation served to relate the set of the Pleiades with the position of the Sun on the horizon. The rise of the Pleiades could then have been observed from the alignment of the west wall of the Coricancha.

6.12 Ceque System and the Stars

As discussed in Chap. 4, the Incas' ceque system surrounding Cusco was extensive. The 41 ceques were filled with huacas for worship and care, at least 328 of them likely being maintained by panacas and ayllus as part of a state-promoted responsibility.

Zuidema argues certain ceques as being straight with intentional astronomical orientations for marking the rising and setting of certain stars. Among these are ceque alignments with such as Betelgeuse, the Pleiades, Vega, and β Centauri. The latter could be used to establish the southern direction (Bauer and Dearborn 1995, p. 139; Niles 1987, p. 176; Zuidema 1977, pp. 250–258; Zuidema 1982a, pp. 219–224; Zuidema 1990a, p. 75).

6.13 Cosmology and Atmospheric Phenomena

The Incas viewed the three worlds of their cosmology to be intertwined by supernatural manifestations. This could be such as the emergence of snakes from the underworld beneath, or in the form of a rainbow from the world above. Many aspects of nature were thought to be sacred and spiritual, such as mountains, rocks, rivers, and caves. Water, in the concept of camay, was thought to be a life-energizing force that was used in many ways to provide life to the inanimate or renew power in the living. Water might charge a huaca with sentience, or be cycled through the heavens, as in its journey down the Vilcanota with return via the Milky Way. The Inca often did not make clear distinctions between celestial and atmospheric phenomena and viewed them all as related to the gods above. The following are occurrences of the atmosphere or seldom recurring astronomical events that helped fill Inca cosmology.

Rainbows

Rainbows were likened with serpents that emerged from springs in the Earth and the underworld. Urton (1981a) says that the Incas believed the rainbow serpents to rise from one spring and descend into another. Serpents were identified with the rainy season, and rainbows, of course, are associated with rain. The rainy season is also the time when the dark constellation Machacuay is most prominent. In Quechua rainbows are referred to as *k'uychi* (Urton 1981a).

Solar and Lunar Haloes

Urton (1981a) says that solar haloes were referred to as *intita chimpusahan k'uychi* and lunar haloes as *quillata chimpushan k'uychi*. K'uychi became part of these terms because halos, like rainbows, were also colored atmospheric phenomena. Haloes were used to predict rain, based on the size of the halo. Larger haloes foretold of impending rain, and smaller ones of rain in the future (Urton 1981a).

Lightning

Zuidema (1964) related that thunder was the messenger of the Sun. In Quechua lightning is referred to as *illapa*. In Urton's study (1981a) he finds it can be either male or female. Female lightning was thought a demon and was of the variety that arrives without sound. It was also said to have the power to harm shepherds who were women, but not their male counterparts. Male lightning came with loud thunder but did not strike the ground.

Meteors

The Quechua word for shooting star is *ch'aska plata*, or "silver star." Meteors were used at times by priests for divination. Shooting stars were said to point to thieves, and also foretell of illness or death (Urton 1981a).

Comets

Comets inspired both awe and fear in the Incas. Many of the celestial bodies of the Incas' cosmos centered on the band of the Milky Way, but a comet could appear anywhere in the night sky and moved from day to day with an ominous tail. Bauer and Dearborn (1995, pp. 148–151, 157) related that Cieza de Leon "suggests that comets were seen on the eve of Atahualpa's death as well as on that of his father, Huayna Capac."

Eclipses
Eclipses could be frightening occurrences for many in ancient civilizations. The Incas worshipped the Sun and Moon as preeminent ancestors and, as such, their sudden disappearances were unnatural events that elicited uncertainty and apprehension. Solar eclipses in a particular region are rare, 17 occurred in the Inca Empire during the reigns of Pachacuti, Topa Inca, and Huayna Capac, including one very prominent eclipse near Cusco in 1513 (Bauer and Dearborn 1995, p. 143). While a solar eclipse may not be visible again in the same location for hundreds of years, lunar eclipses are visible over wide areas and are therefore relatively common. Incas believed that when the Sun eclipsed, he was angry and when the Moon, did so she was ill, and they would wail. Following a darkening of the Sun rituals were performed, and sacrifices were made of livestock and children. With the loss of the Moon dogs were tied and beaten so that they would howl and bring her back (Cobo 1990 [1653], p. 27).

Twilight
Andean cosmology views twilight as a region of space that trailed the Sun in the evening and preceded it in the morning, more so than simply a period of time. This "space" was important, as it was where the heliacal rises of certain celestial bodies were witnessed (Urton 1981a, pp. 151–153). While not proven, this may be a long-standing tradition.

6.14 Summary

Many of the huacas and temples that remain in the former Inca Empire are testaments to the sophisticated celestial system that developed in the Andes. Cosmology began with worship of the Sun, and to a lesser extent the Moon, with the state promoting the ruling Inca as the "son of the Sun." With such basic tenets to build upon, it was inevitable that the Incas would adopt and develop an astronomy of a significant solar nature.

Their celestial interests were much broader, however, and included many other objects in the heavens. The axis of the Milky Way is very prominent at the latitude of Cusco and the Andean peoples adopted it as their primary reference in the night sky. While European constellations evolved around the ecliptic path of the Sun, celestial images in the Inca Empire remained within, or close to, and ordered by the brilliant band of our galaxy.

The Moon played a significant role, but due to its female connotations those functions are far less documented when compared with the masculine Sun. The Incas used both the Sun and the Moon in the regulation of their year (Zuidema 2007).

The calendar of Cusco was publicly observed by solar pillars constructed on the city's horizons. A sophisticated horizon astronomy developed where such towers assisted in monitoring the annual path of sunrises and sunsets. The pillars of Cusco were destroyed in a Spanish purge of idolatry and varying accounts disagree as to their numbers and placement. It is likely that they marked the times of the December and June solstices and as well the days of the zenith and anti-zenith Sun. Away from Cusco, two towers still exist above the modern village of Urubamba and are examined in detail in Sect. 8.3. Solar events also were likely observed with reference to the natural horizon in areas where those features were significant.

Evidence of the importance of astronomy in Inca culture can be seen in the many celestial alignments to be found at temples and in extant huacas of carved rock. Many of these are examined in Chaps. 7, 8, and 9.

Zuidema proposed an astronomically based calendar using 328 huacas on 41 ceques surrounding Cusco. The remaining 37 days of the 365-day tropical solar year were represented by the approximate period of the annual disappearance of the Pleiades behind the Sun.

The Pleiades were worshipped by the Incas and played a major role in society as a harbinger of maize production. While being unaware of the true nature of El Niño, the Andeans had discovered a means by which to predict its imminence. Orlove, Chiang, and Cane (2000, p. 68) argue that the degree of brilliance and visibility of the stars of the Pleiades following their heliacal rise in June bode for a bountiful crop in normal years, or a depleted one with El Niño.

Much of Inca cosmology and myth was represented in the system of dark constellations they visualized in the parts of the Milky Way blocked by interstellar dust and gas. The world of the heavens above was closely intertwined with both the terrestrial world and the subterranean underworld. Powerful and influential spirits lived in many features of the natural worlds and also in the cosmos. The entities of these dark constellations figured prominently in everyday life.

Inca astronomy permeated society and culture. The state purposely created numerous constructs with astral orientations, both to take advantage of the regulatory nature of recurring celestial events and also to create an aura of connectivity with the heavens in an effort to further establish power and legitimacy. The astronomical alignments that remain in huacas and temples today are testaments to the sophistication of the system that evolved. The many celestial orientations found in the Inca Empire can only be fully understood when taken in context with the Incan astronomy described in this chapter, an astronomy that was founded upon principles in a frame of reference different from that of our own.

Chapter 7
Orientations at or near Cusco, at Tipon, and at Saihuite

Fig. 7.1 The Saihuite Principal Stone (watercolor by Jessica Gullberg)

Pachacuti instituted a great system of ceques and huacas that emanated in all directions from Cusco's Coricancha, the primary Sun temple of the Incas. Pachacuti ordered many Sun temples to be built across the empire so that all could worship the Sun (Cobo 1983 [1653], p. 134). Structures such as the Coricancha, the Sunturhuasi, and the ushnu of the Hanan Huaycapata plaza figured prominently in Inca astronomy and are discussed in Chap. 6. At least 328 huacas were worshipped and cared for, but most did not survive the Catholic purge of idolatries following

© Springer Nature Switzerland AG 2020

S. R. Gullberg, *Astronomy of the Inca Empire*, Historical & Cultural Astronomy,
https://doi.org/10.1007/978-3-030-48366-1_7

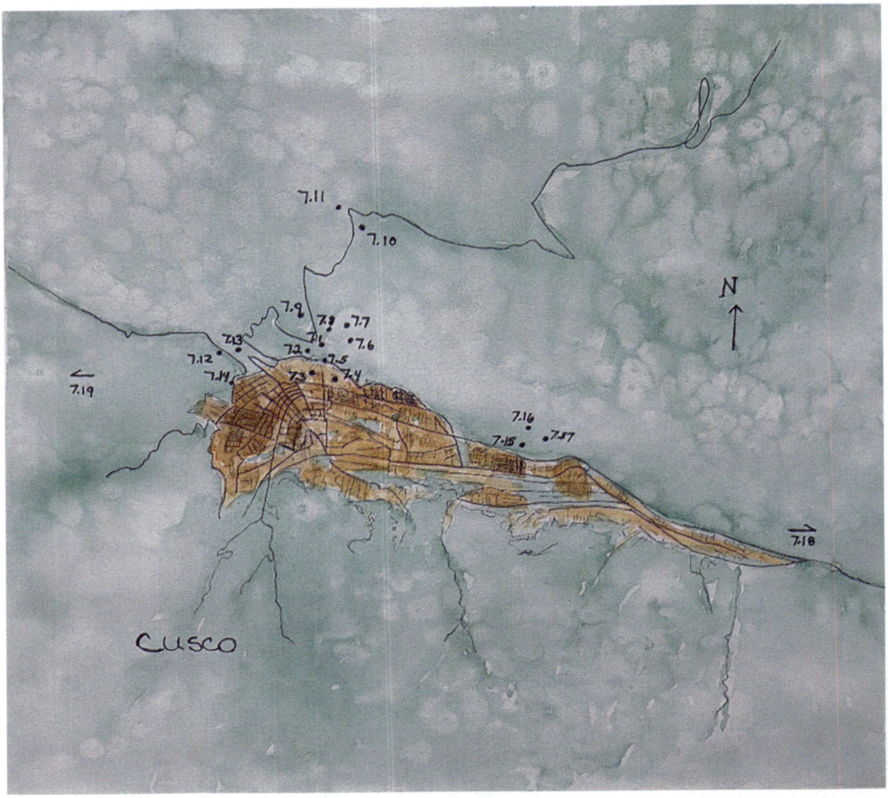

Fig. 7.2 Sites profiled in Chap. 7 (watercolor map by Jessica Gullberg). 7.1 Kenko Grande*. 7.2 Kenko Chico*. 7.3 Mesa Redonda. 7.4 Tetecaca. 7.5 Patallacta. 7.6 Kusilluchayoc*. 7.7 Lacco*. 7.8 Huaca 44*. 7.9 Lanlakuyok*. 7.10 Puca Pucara*. 7.11 Tambomachay*. 7.12 Sacsahuaman*. 7.13 Mollaguanca*. 7.14 Sapatiana. 7.15 Rumiwasi Bajo*. 7.16 Rumiwasi Alto*. 7.17 Kusicallanca*. 7.18 Tipon*. 7.19 Saihuite*. *Astronomical orientation identified

the Spanish conquest. Those that remain largely were carved from rock and many of these are examined. The system for designating ceques and huacas that was first mentioned by Cobo (1990 [1653], pp. 51–84) and later put forth by Bauer (1998) is used here. Certain huaca descriptions by Cobo are included. Featured are photographs of astronomical effects found with many of the carved rock huacas, stones, caves, and structures. Also included are negative findings for certain huacas and carved rocks that exhibited no signs of celestial orientation. Calculated azimuths were arrived at with spherical trigonometry (Sect. 5.7; Appendix 6). Figure 7.2 is a location map of the sites presented in this chapter, including those north, east, and west of Cusco.

7.1 Kenko Grande

Visible from the Coricancha, Kenko Grande (Fig. 7.3) is a sculpted limestone outcrop north of Cusco that incorporates carvings and crevices as well as a figurative puma-shaped stone monolith surrounded by a series of niches along an arced plaza. It is located at S13°-30.53′; W071°-58.24′ and is at 3614 masl. Kenko Grande is a carved outcropping and was found to exhibit light and shadow effects, solsticial orientations, equinoctial orientations, stairs, seats, niches, a water source, basins, a cave, a light-tube, altars, a monolith, animal replica stones, and a zigzag channel.

The Awakening of the Puma

Kenko Grande exhibits a dramatic phenomenon at the time of the June solstice that is known locally as "the awakening of the puma." The Inca venerated the condor, puma, and snake with regard to their cosmological correlations with the sky, Earth, and underworld (Van de Guchte 1990). Located atop the huaca and carved into the stone are two round gnomons designed for effects of light and shadow (Fig. 7.4). The gnomons are about 25 cm high and are spaced 35 cm apart. In close proximity is a small wall with a fissure (Figs. 7.5 and 7.7) that is aligned for sunrise at June solstice. Light from the morning Sun passes through the fissure and first touches the left side of the left gnomon. As the Sun continues to rise its rays move across the gnomon and then illuminate the opposite one as well. The gnomons are situated in

Fig. 7.3 Kenko Grande and its monolith (author's photograph)

Fig. 7.4 Gnomons carved on top of Kenko Grande (author's photograph)

Fig. 7.5 Fissure aligned for the June solstice sunrise (author's photograph)

such a way that the glowing pair and the associated shadows now resemble a puma – "the puma's awakening" (Figs. 7.6 and 7.7).

Light falling on the gnomons is thought by some to help mark the equinoxes (Fig. 7.8) and the December solstice. On the day of an equinox the Sun rises between

Fig. 7.6 The Eyes of the Puma at June solstice sunrise (author's photograph)

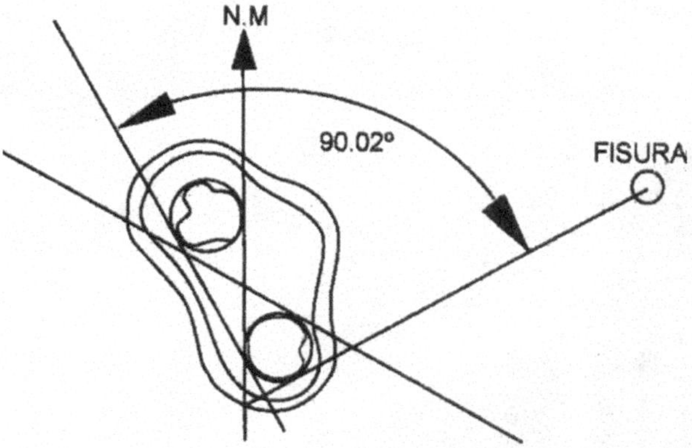

Fig. 7.7 Plan of the gnomons and fissure. The line of the fissure (fisura) in the drawing is 62° from north (from Arévalo 2007)

the gnomons on the eastern horizon. Its rays soon cast an effect of light and shadow that splits the carving into dark and light halves, said to represent the equal periods of day and night. The December solstice Sun rises over the 6372-meter snowpeak Nevado Ausangate with an alignment allowing sunlight to cross between and be tangent to both gnomons. There is not sufficient evidence to support December solstice and equinox use, but the Incas did at times create sites for multiple purposes. The puma effect at Kenko Grande is well known and observed by many during the June solstice sunrise.

Fig. 7.8 According to local folklore, light and shadow effects at equinox represent the division of day and night (author's photograph)

Fig. 7.9 The zigzag channel carved on top of Kenko Grande (author's photograph)

Zigzag channel

The top of Kenko Grande exhibits many once-fine carvings that include an offertory/divination channel flowing toward a cave (Fig. 7.9). A common motif of carved huacas is a straight or zigzag channel through which liquids, possibly chicha (Fig. 3.23), could flow. The current of energy necessary to establish harmony and maintain equilibrium in the world was stimulated by the pouring of liquid offerings

Fig. 7.10 Kenko Grande's primary altar (author's photograph)

into these channels. This channel begins with a cup-like basin, extending first for 130 cm and then 145 cm more before splitting into two 80 cm branches. *Kenko* in Quechua means a zigzag (Paternosto 1989).

Interior cave and altars

Located within Kenko Grande is a sculpturally enhanced cave with two entrances, two altars, niches, and a set of three ritual stairs (Fig. 7.10). Locals maintain that light entering the cave within the niche shown at the rear of this photo was reflected by the Incas with gold or silver plates in order to illuminate the entire chamber. The cave is oriented along the solsticial southeast-northwest axis and includes an additional niche to the right of the altar large enough to hold a mummy.

The primary altar was carved and polished with three ritual stairs at its northwestern end (Fig. 7.11). During the time about local noon on days nearing the June solstice, sunlight enters the cave, approaches the altar, and climbs the three stairs, adding one at a time beginning with the bottom step (Figs. 7.12 and 7.13). Ritual stairways are common features symbolizing shamanic movement between the three cosmological worlds. Such a cave would serve as a portal for communication with Ucu Pacha, the world below. Light enters through an opening approximately 341.5° in azimuth and inclined by 40°.

At times of the solar equinoxes, the cave's secondary altar is illuminated at sunrise (Fig. 7.14). The opposite entrance of the cave near the main altar is oriented to the equinox sunset. The meaning of this remains uncertain as ethnohistorical evidence has not been found to fully describe equinox activities.

Fig. 7.11 Left side of the altar with steps as sunlight approaches (author's photograph)

Fig. 7.12 Sunlight has climbed the first two steps (author's photograph)

Fig. 7.13 Midday June solstice sunlight has climbed all three symbolic steps and points toward the altar (author's photograph)

Fig. 7.14 The secondary altar within Kenko Grande's cave at equinox sunrise (author's photograph)

Fig. 7.15 The monolith at Kenko Grande (author's photograph)

Monolith

Kenko Grande's 5.5-meter-high monolith (Fig. 7.15) is located near the focal point of a semicircular shaped plaza that includes a partially destroyed wall containing 19 body-sized niches, plus a 20th opening for an entranceway. Modern interpretations list this as a puma, a phallic symbol, or an *axis mundi*, and the monolith may also have functioned as a gnomon.

7.2 Kenko Chico

Kenko Chico (Fig. 7.16) is near Kenko Grande to its southwest, separated by a ravine and a small stream. It is located at S13°-30.62′; W071°-58.32′ and at 3598 masl. Kenko Chico is a carved rock and was found to exhibit a solsticial orientation, a water source, basins, and a gnomon. It contains a stone stairway leading to its summit, terraces, and large stone blocks similar in size to many of those used at Sacsahuaman. It is surrounded by a moat and exhibits many carvings, such as steps and seats. Kenko Chico is oriented generally east-west and, while not a structural alignment, Nevado Ausangate can be viewed on the horizon in the general direction of the December solstice sunrise. The central staircase is aligned north and south.

Fig. 7.16 Kenko Chico (author's photograph)

7.3 Mesa Redonda

In the residential area immediately below Kenko Chico and visible from the Coricancha is a site known as Mesa Redonda (Fig. 7.17). Mesa Redonda is located at S13°-30.80′; W071°-58.18′ and is at 3474 masl. It is a group of carved rocks and was found to exhibit stairs, a water source, and an altar. Bauer (1998) states that the boulder and adjacent carved rock are likely candidates for Pachatosa (An 2:2).

> [An 2:2] the second *guaca* of this *ceque* was named Pachatosa; it was a large stone which was next to [Diego] Cayo's house. The sacrifice was burned on top of it, and they said that the stone ate it. (Cobo 1990 [1653], p. 64)

7.4 Tetecaca

To the east of Mesa Redonda and also visible from the Coricancha is a large carved outcrop known locally as Tetecaca (Fig. 7.18). Tetecaca is located at S13°-30.90′; W071°-57.93′ and is at 3492 masl. It is a carved rock and was found to exhibit a niche, a water source, a cave, an altar, and a channel.

Fig. 7.17 Mesa Redonda (author's photograph)

Fig. 7.18 Tetecaca (author's photograph)

The boulder includes a modern shrine and has a gnomon sculpted upon its upper surface. The site has a good view of the valley and surrounding horizons and the top of the rock includes a mini-landscape somewhat reminiscent of the one found at Saihuite.

7.5 Patallacta

Van de Guchte (1990) calls this outcrop Patallacta (Ch 1:2). Cobo describes a house that may have been nearby.

> [Ch 1:2] The second *guaca* of the *ceque* was called Patallacta. It was a house which Inca Yupanqui designated for his sacrifices, and he died in it. The Incas who succeeded him thereafter made ordinary sacrifice here. In general, all the things which they consumed in sacrifice were offered for the health and prosperity of the Inca. (Cobo 1990 [1653], p. 51)

Patallacta is located at S13°-30.75′; W071°-58.11′ and at 3492 masl. It is a carved rock that is visible from the Coricancha and has been damaged by the modern road cut below it (Fig. 7.19). Carvings, especially those of seats, are extant and from the top of the rock there is a commanding view of the north half of the valley where horizon events could have been observed.

Fig. 7.19 Patallacta (author's photograph)

Fig. 7.20 The western face of Kusilluchayoc (author's photograph)

7.6 Kusilluchayoc

Proceeding upslope from Patallacta is another large limestone outcrop, Kusilluchayoc, which is situated about 300 meters south of Lacco (Fig. 7.20). It is located at S13°-30.51′; W071°-57.94′ and is at an elevation of 3637 masl.

Kusilluchayoc is a carved rock with adjacent structures and was found to exhibit a solsticial orientation, niches, a water source, a basin, animal replica stones, and a channel. In Quechua *Kusilluchayoc* means "the one with the monkey" (Van de Guchte 1990) and among the carvings on the huaca are those of three monkeys, a serpent, and a puma.

Several masonry walls form multiple rooms with windows and niches on Kusilluchayoc's western face. Both a corridor between the rocks and a body-size niche face the setting point of the June solstice Sun (Fig. 7.21).

7.7 Lacco

Lacco is a carved limestone outcropping incorporating several astronomical alignments (Fig. 7.22) that was barely visible from the Coricancha. Lacco is located at S13°-30.29′; W071°-57.89′ and at 3662 masl. It was found to exhibit light and shadow effects, a solsticial orientation, a zenith orientation, stairs, seats, niches, a

Fig. 7.21 View from Kusilluchayoc in the direction of the June solstice sunset (author's photograph)

Fig. 7.22 Lacco with Nevado Ausangate in the distance (author's photograph)

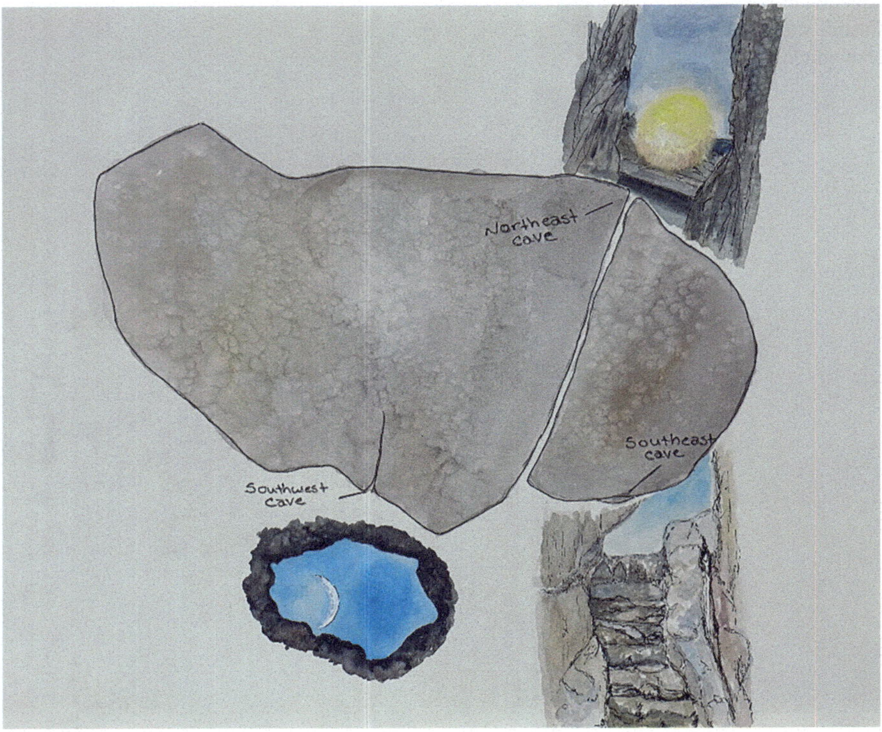

Fig. 7.23 Plan of Lacco. The Northeast cave is oriented for direction of the June solstice sunrise. The Southwest and Southeast caves contain light-tubes (watercolor by Jessica Gullberg)

water source, basins, caves, light-tubes, altars, gnomons, animal replica stones, and channels (Fig. 7.23). The elaborate huaca, also known as Salonpuncu, is above Cusco to the northeast of Kenko Grande. Zuidema (1977) and Aveni (1981a) suggest Lacco as near the site of Chuquimarca (An 3:4).

[An 3:4] The fourth was called Chuquimarca; it was a temple of the Sun on the hill of Manto calla, in which they said that the Sun descended many times to sleep. For this reason, in addition to everything else, they offered it children. (Cobo 1990 [1653], p. 65)

Lacco is the largest limestone outcrop in the vicinity, covering an area of some 1670 square meters. There are at least seven other carved outcrops located within a 1 km radius. Two stairways face northwest toward the June solstice sunset. Lacco's upper surface is extensively carved with examples of such as seats, sacred animals, and a gnomon. A crevasse runs like a corridor from the northeast side to the southeast, where there are short sections of stairways and a series of niches large enough to stand inside. A canal, originating at Ucu Ucu some 3 kms distant, runs near the base of the rock and the Inca road to Pisac is nearby as well. Three caves were modified with altars and exhibit astronomical orientations.

Fig. 7.24 Lacco's Northeast cave and two ceremonial thrones situated before it (author's photograph)

Northeast cave

Lacco, on its northeast face, has a cave opening oriented for the June solstice sunrise (Fig. 7.24). The greatest eastward angle as viewed through the cave's entrance is 078°, and at least some sunlight is admitted to a portion of the interior for several days before and after the solar standstill. The Sun centers on the cave opening at the time of the June solstice (Fig. 7.25). During this period sunlight enters the portal, illuminating the altar and cave interior (Fig. 7.26). The process observed began very quickly at 06:25 and persisted until the last vestiges of light disappeared from the altar's stone surface nearly two hours later at 08:24 (Fig. 7.27). The Northeast cave is located at S13°-30.34′; W071°-57.86′ and 3650 masl.

Immediately to the east of the Northeast cave is the north opening of the crevasse running across Lacco. It also is oriented approximately to the June solstice sunrise. Two ceremonial thrones are situated in front of and below two large steps leading to the Northeast cave's opening. The thrones might have been occupied during the rising of the solstice Sun.

Southwest cave

Lacco's Southwest cave contains a small altar (Fig. 7.28) below a light-tube oriented to the area of the ecliptic at times when the Sun's path is between 70° and 75° above the horizon (Fig. 7.29). When properly positioned, either the Sun or a brilliant

Fig. 7.25 June solstice sunrise as seen from Lacco's Northeast cave. The Sun advances each day from the right until reaching the central standstill as depicted, after which it reverses course and retreats daily to the right again (author's photograph)

Fig. 7.26 Lacco's Northeast cave and altar illuminated by the June solstice early morning Sun (author's photograph)

Fig. 7.27 As the Sun continues to rise, the area of its light cast upon the altar gradually grows smaller until it completely disappears (author's photograph)

Fig. 7.28 Altar in the southwest cave of Lacco (author's photograph)

Moon can illuminate the altar within. A crescent Moon was viewed through the light-tube near the October 30th date of the zenith Sun (Fig. 7.30). The cave's door is oriented facing 235°. The Southwest cave is located at S13°-30.35′; W071°-57.89′ and at 3662 masl.

Fig. 7.29 Looking up through the light-tube (author's photograph)

Fig. 7.30 The crescent Moon as viewed upward through the light-tube in Lacco's Southwest cave (author's photograph)

The altar within the cave is aligned to be illuminated by light passing through the tube. The inclination from the altar to the lower edge of the light-tube is approximately 70° at an azimuth of 211°. The cave includes two recesses cut into its western wall.

Fig. 7.31 The entrance to Lacco's Southeast cave (author's photograph)

Southeast cave

Lacco's second cave with light-tube illumination is known as the Temple of the Moon (personal communication, Milton Rojas Gamara 2017) but is also illuminated by the Sun. The chambers are the most elaborate found within the huaca and the remains of carvings of both a puma and a snake adorn the entrance (Figs. 7.31 and 7.32), the serpent symbolizing passage into the realm of Ucu Pacha (Paternosto 1989). Sunlight is admitted through the cave's light-tube near solar noon at the time of the zenith Sun. This observance took place with the Sun at an inclination of 89° (Fig. 7.33). The light-tube is directed at a finely carved altar (Fig. 7.34) on an azimuth of 055°, which can also be illuminated by the Moon. The Southeast Cave is located at S13°-30.36′; W071°-57.86′ and at 3655 masl.

7.8 Huaca 44

Located between Kenko Grande and Lacco (Fig. 7.35) is a small often overlooked huaca exhibiting interesting alignments for the solstices and equinoxes (Fig. 7.36). This huaca indicating the direction of primary solar horizon events is known as

Fig. 7.32 A carved snake adorns the entrance of Lacco's Southeast cave (author's photograph)

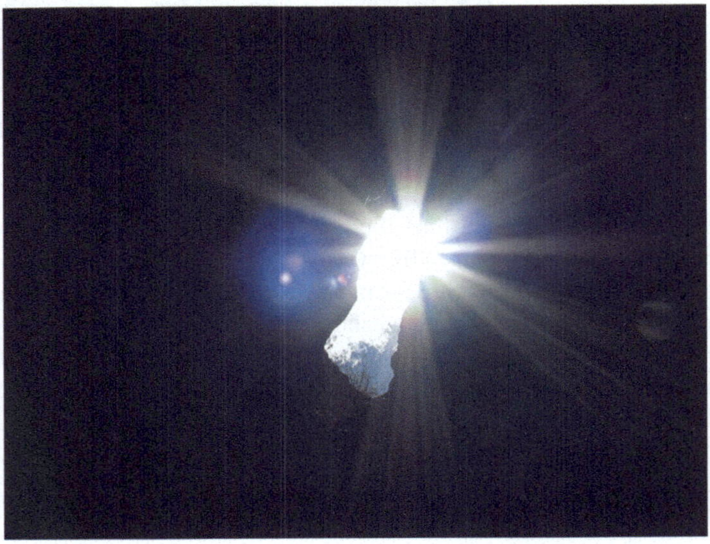

Fig. 7.33 The Sun as viewed through Lacco's Southeast cave's light-tube (author's photograph)

Huaca 44 (personal communication, Milton Rojas Gamara 2019; he performed some of the early research at the site). Within it are two circular flat carvings, one large and one small, oriented in such a way that they could have had utility as solar gauges (Fig. 7.37). A line drawn across the circular carvings, aligned east-west with respect to one another, would indicate the horizon positions of the

Fig. 7.34 The altar illuminated by the zenith Sun (author's photograph)

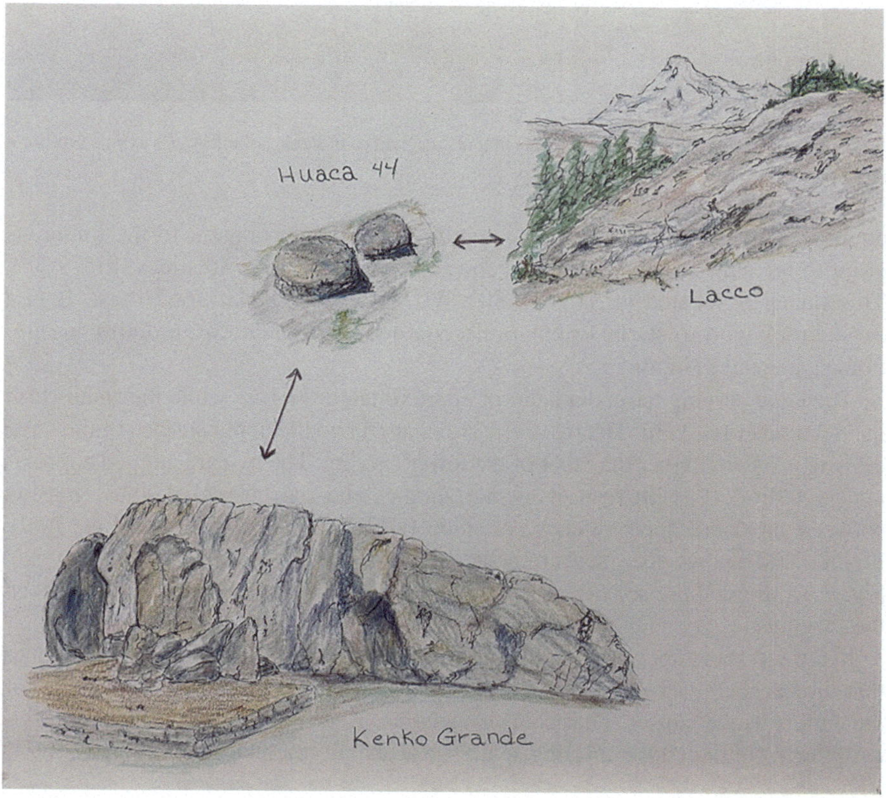

Fig. 7.35 Relative position of Huaca 44, the huaca with gnomons aligned to indicate the direction of primary solar horizon events (watercolor map by Jessica Gullberg)

Fig. 7.36 Huaca 44 and its two gnomons that are aligned to indicate solar horizon events (author's photograph)

equinox sunrises and sunsets. Intersecting lines drawn tangent to the gnomons point to the sunrises and sunsets of the June and December solstices (Fig. 7.38). This huaca is located at S13°-30.30′; W071°-58.07′ and at 3662 masl. It is a carved rock with solsticial orientations, equinoctial orientations, a zenith orientation, seats, and gnomons.

The large carving has a diameter of approximately 110 cm while the small carving's diameter is 43 cm. The two are 131 cm apart and the upper surface of the large carving is 55 cm higher than that of the lower carving. The two are aligned approximately 090°/270° with respect to one another and the tangential lines crossing between them are approximately oriented 065°/245° and 114°/294° (Fig. 7.39). Figure 7.40 shows the measurement for June solstice sunrise orientation and Fig. 7.41 shows June solstice sunrise as viewed along the tangential line between the carvings.

Seats and stairs are carved in the huaca and several seats are aligned in a manner that directs them toward viewing these solar horizon events. The exception is for the June solstice sunset. That area of the stone is severely eroded but may originally have contained the missing seat. The seats are oriented to 063°, 092°, 116°, 247°, and 272°.

The overall purpose of this shrine remains enigmatic. While astronomical alignments are present, it is possible that the Incas used the site for other purposes

Fig. 7.37 A large and a small gnomon measured by the author's son, Greg, are carved into the huaca's base (author's photograph)

as well. The two carvings might have served as ceremonial platforms, giving rise to the possibility of an additional astronomical function when the day of the zenith Sun could be determined with a figure placed on one of the carvings casting no shadow.

7.9 Lanlakuyok

Lanlakuyok is a large carved outcrop along the road to Pisac that has an extensive system of passageways within it and is illuminated by occasional tubes or openings to the sky (Fig. 7.42). Lanlakuyok is located at S13°-29.82′; W071°-58.41′ and at 3727 masl. It was found to exhibit a solsticial alignment, an equinoctial orientation, stairs, seats, niches, a water source, caves, and light-tubes (Fig. 7.43). The geographical position of the huaca allows a clear view of both Lacco and Sacsahuaman (Fig. 7.44). Kenko Grande could also have been

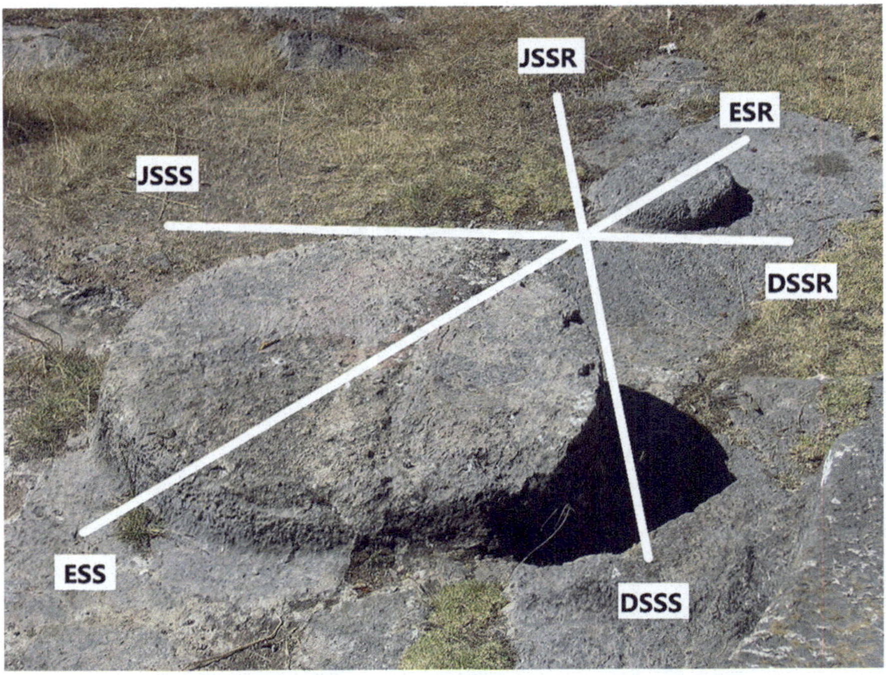

Fig. 7.38 Gnomon alignments indicate the directions of the horizon positions of sunrise and sunset on days of the solstices and equinoxes (from author's photograph)

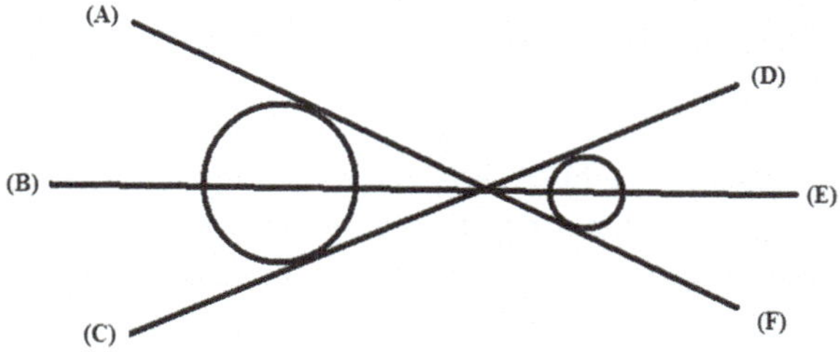

Fig. 7.39 Orientation of the gnomons for six solar horizon events. Line B-E is oriented for the equinox sunrise and sunset. Line C-D is oriented for the June solstice sunrise and the December solstice sunset. Line A-F is oriented for the December solstice sunrise and the June solstice sunset (author's diagram)

visible but presently is blocked by trees. A primary cave opening faces the equinox sunrise.

Sherbondy and Van de Guchte have suggested that Lanlakuyok is a candidate for Amaromarcaguaci (An 1:7).

> [An 1:7] The seventh *guaca* was called Amaromarcaguaci; this was a house of Amaro Tupa Inca, which was on the road of the Andes. (Cobo 1990 [1653], p. 63]

Fig. 7.40 Initial measurement for June solstice sunrise orientation (author's photograph)

Fig. 7.41 June solstice sunrise as viewed between the gnomons (author's photograph)

7.10 Puca Pucara

Puca Pucara is said to have been one of Pachacuti's residences (Bauer 1998) and Niles (1987) has suggested it as the location of An 1:9.

Fig. 7.42 Entrance to Lanlakuyok (author's photograph)

> [An 1:9] The ninth was named Tambomachay; it was a house of Inca Yupanqui where he lodged when hunting. It was set on a hill near the road of the Andes. They sacrificed all kinds of things to it except children. (Cobo, 1990 [1653], p. 63]

Puca Pucara is located at S13°-29.00′; W071°-57.72′ and is at 3805 masl. The huaca is located at the top of a small hill along the road to Pisac and consists of several structures and a fountain. Entry to the main part of the compound is through a pair of double-jambed doorways connected by a corridor. The doors and corridor are oriented approximately 090°/270° on an axis that also aligns with the fountain (Fig. 7.45). The group of structures were found to exhibit cardinal and equinoctial orientations, as well as a water source and platforms.

Bauer (1998) states that the fountain at Puca Pucara (Fig. 7.46) should be considered as a candidate for Puquiu (An 2:8).

> [An 2:8] The eighth *guaca* was called Puqui; it was a fountain which is at the end of Tambo Machay. They offered it sheep, clothing, and shells. (Cobo, 1990 [1653], p. 64)

7.11 Tambomachay

The cave of Tambomachay (Fig. 7.47) is associated with an Inca platform and staircase and Bauer (1998) suggests it to be the best candidate for Cirocaya (Ch 1:4).

Fig. 7.43 Author recording data at a naturally lit point in a passageway within Lanlakuyok (author's photograph)

> [Ch 1:4] The fourth *guaca* was called Cirocaya. It is a cave of stone from which they believed the hail issued. Hence, at the season when they were afraid of it, all went to sacrifice in the cave so that hail should not come out and destroy their crops. (Cobo 1990 [1653], p. 54)

The cave opening looks out on a bearing of 135° while the platform more directly faces the December solstice sunrise. The mouth of the cave is located at S13°-28.58′; W071°-58.09′ and is at 3900 masl. The site is positioned with a commanding view of Tambomachay, Puca Pucara, and parts of the Cusco Valley. The cave of Tambomachay is part of a rock outcrop and was found to exhibit a solsticial orientation. There is water source at the Fountain of Tambomachay below and a stairway leads to the south end of the platform. A boulder rests centrally on the terrace of the platform and a wall was constructed on the side of the terrace closest to the outcrop. The cave is shallow, and its interior is extensively eroded.

Bauer (1998) proposes that the highly structured fountain at Tambomachay (Fig. 7.48) may have been Quinoapuquiu (An 1:10).

Fig. 7.44 Sacsahuaman's zigzagged walls as viewed from Lanlakuyok (author's photograph)

Fig. 7.45 Double–jambed doorways and corridor at Puca Pacara oriented east-west (author's photograph)

Fig. 7.46 A fountain is aligned on the east-west axis (author's photograph)

Fig. 7.47 The cave of Tambomachay with a platform oriented to the December solstice sunrise (author's photograph)

Fig. 7.48 The fountain of Tambomachay (author's photograph)

> [An 1:10] The tenth *guaca* was called Quinoapuquiu; it was a fountain near Tambo Machay which consists of two springs. Universal sacrifice was made to it, except children. (Cobo 1990 [1653], p. 64)

The fountain is situated on the main level of the complex, well below the cave and platform.

7.12 Sacsahuaman

Niles (1999) feels that construction of Sacsahuaman (Fig. 7.49) may have been started by Topa Inca and completed by Huayna Capac. Thousands may have worked for over 50 years on its construction (Salazar and Salazar 2014). The base of Sacsahuaman's zigzag walls are located at S13°-30.50′; W071°-58.91′ and at 3595 masl. Zuidema and Quispe (1973) state that Sacsahuaman was regarded as a temple of the Sun. The orientation of Cusco with Sacsahuaman northwest and above it ensured the June solstice Sun would first strike Sacsayhuaman before illuminating the city. Zuidema said that a seat in Sacsahuaman is on a northwest/southeast line running from Ollantaytambo to Cusco and on to Lake Titicaca (personal communication 2007). Inti Raymi is celebrated on the field of Sacsahuaman in June, and in

Fig. 7.49 Plan of Sacsahuaman as viewed from the south with the zigzagged walls across the top and Muyucmarca left of center (watercolor by Jessica Gullberg)

September as well is the Festival of Warachicuy, a series of tests of prowess and rites of passage for boys to enter manhood (Salazar and Salazar 2014). Sacsahuaman is a multifaceted complex that exhibits carved rocks, zenith and anti-zenith alignments, stairs, seats, niches, a water source, basins, a cave, and altars.

Muyucmarca

Above the zigzagged walls of Sacsahuaman (Fig. 7.52) lies the foundation of a round building called Muyucmarca (Figs. 7.50 and 7.51), which Zuidema describes as having three concentric round walls divided by interior walls and three gutters. He states that the central of three water channels, its opposing niche, and the walls that parallel them are all aligned along the axis of the zenith sunrise and anti-zenith sunset (Fig. 7.52). Muyucmarca's round tower and related buildings could have been used to observe the zenith Sun overhead, as well as the horizon positions of the sunrise at zenith and the sunset at anti-zenith (Zuidema 1981b, p. 326).

Tired Stone

The Tired Stone (Fig. 7.53) is Piedra Casada (Ch 4:6).

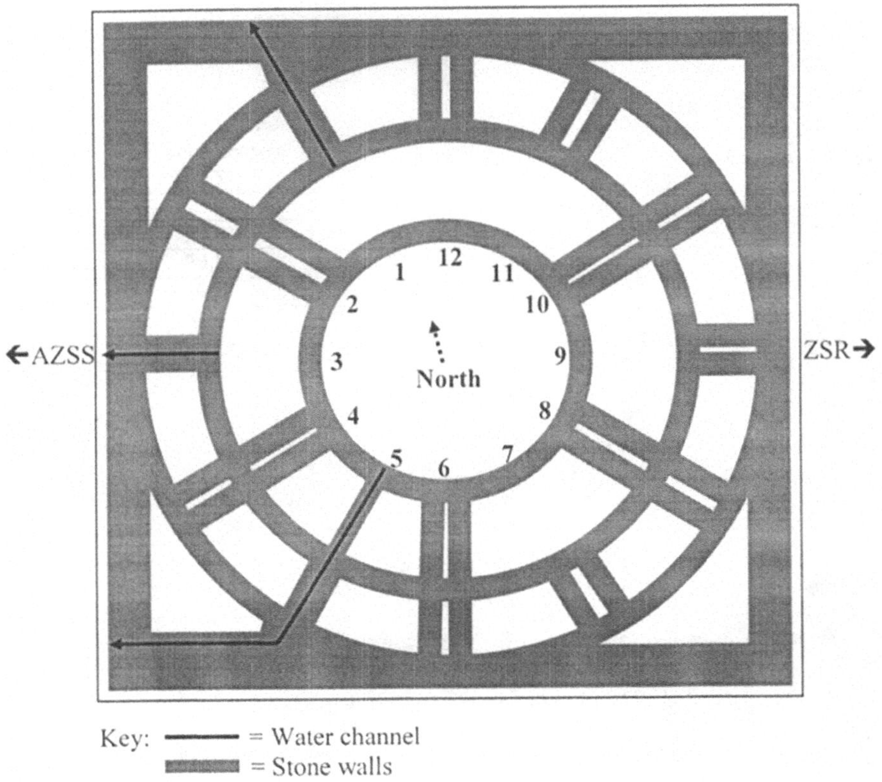

Fig. 7.50 The zenith and anti-zenith alignments of Muyucmarca (from Zawaski 2007, p. 58)

> [Ch 4.6] The sixth *guaca* was a large stone named Collanconcho which was in the fortress. They declared that, bringing it for that structure, it fell three times and killed some Indians. The sorcerers said that it had replied to questions they put to it that, if they persisted in wanting to put it in the structure, all would have a bad end, apart from the fact that they would not be able to do it. From that time on it was considered a general guaca to which they made offerings for the strength of the Inca. (Cobo 1990 [1653], pp. 56–57)

This limestone huaca is extensively carved with many examples of seats and stairs. Legend has it that as the massive stone was being brought to Sacsahuaman it became "tired" and could move no further (Hemming and Ranney 1982). Van de Guchte (1990) argues the exclusion of the Tired Stone and the subsequent incompletion of Sacsahuaman to be intentional. This would leave the project's completion for supernatural assistance. The Tired Stone is located at S13°-30.25′; W071°-58.86′ and at 3610 masl.

Throne of the Inca

Across the plaza from Sacsahuaman's ramparts is a huaca known as the Throne of the Inca (Fig. 7.54), or Sabacurinca (Ch 5:6).

Fig. 7.51 Muyucmarca (author's photograph)

Fig. 7.52 Viewed from the north, Muyucmarca is to the right of center above the zigzagged walls of Sacsahuaman (author's photograph)

Fig. 7.53 The Tired Stone (author's photograph)

Fig. 7.54 The Throne of the Inca (author's photograph)

[Ch 5:6] The sixth *guaca* was named Sabacurinca; it was a well-carved seat where the Incas sat. It was very venerated, and solemn sacrifices were made to it. On account of this seat, the whole fortress was worshipped, for the seat must have been inside or next to the fortress. (Cobo 1990 [1653], p. 57)

The series of massive steps or tiers were carved from diorite and are said to have served as a place where ruling Incas sat, in particular in a seat carefully carved for that purpose. The Throne of the Inca faces Cerro Huanacauri, the mountain mentioned with the *procession* and the *pilgrimage* in Sect. 3.16. It is located at S13°-30.43′; W071°-58.93′ and is at 3618 masl.

7.13 Mollaguanca

To the east of Sacsahuaman lies a small huaca that may be Mollaguanca (Ch 3:6). Mollaguanca is located at S13°-30.35′; W071°-58.74′ and is at 3581 masl.

[Ch 3:6] The sixth *guaca* was named Mollaguanca; it was a certain stone which was in the middle of a flat place which they called Calispuquiu. Inca Yupanqui ordered it to be placed there and considered it a shrine. (Cobo 1990 [1653], p. 55)

Mollaguanca is a carved rock and was found to exhibit a solsticial orientation and a water source. The crevasse in the center of the boulder (Fig. 7.55) is aligned on the axis for June solstice sunrise and December solstice sunset. There are

Fig. 7.55 Mollaguanca (author's photograph)

Fig. 7.56 Sapatiana is now surrounded by homes (author's photograph)

carvings of stairs and seats, but erosion is extensive thus leaving these potential alignments for June solstice sunrise and December solstice sunset inconclusive.

7.14 Sapantiana

In a residential area below Sacsahuaman and within view of the Coricancha is a huaca known as Sapantiana (Fig. 7.56). Sapantiana is located at S13°-30.73′; W071°-58.68′ and is at 3465 masl. It is a carved rock and was found to exhibit a water source and an animal replica stone. Homes have grown to surround this shrine that Bauer (1998, p. 53) says is the "best possible location for Ch 2:4."

> [Ch 2:4] The fourth guaca was called Viroypacha; it is a conduit of fairly good water which was declared a guaca by Inca Yupanqui. It was prayed to for the tranquillity of the Inca. (Cobo 1990 [1653], p. 54)

7.15 Rumiwasi Bajo

To the east of Cusco above San Sebastián lies Rumiwasi Bajo, located at S13°-31.17′; W071°-56.42′ and at 3470 masl. Rumiwasi Bajo is a group of structures and was found to exhibit a solsticial orientation, niches, and a cave (tunnel). The rock of

Fig. 7.57 The southeastern door to the passageway of Rumiwasi Bajo (author's photograph)

Rumiwasi Bajo contains a number of niches and a nine-meter-long passageway. One doorway to the passageway through the huaca looks out in the general direction of June solstice sunset but is 12° off. The other doorway opens to the December solstice sunrise (Fig. 7.57).

7.16 Rumiwasi Alto

Another carved rock lies several hundred meters uphill, known as Rumiwasi Alto. Rumiwasi Alto is located at S13°-30.97′; W071°-56.36′ and at 3563 masl. The site includes four terraces bound by walls, one with four niches facing out to 195° (Fig. 7.58). Several other niches have not survived. The group of structures was found to exhibit a solsticial orientation, niches, and terraces. Rumiwasi Alto looks 13.5° down upon Rumiwasi Bajo at an angle of 206°. The walls of the terraces are aligned at 118° and nearby Kuscallanca is oriented with this site on the December solstice sunrise axis.

Fig. 7.58 Walls and niches of Rumiwasi Alto (author's photograph)

7.17 Kusicallanca

Kusicallanca neighbors Rumiwasi Alto to the southeast. It is located at S13°-31.02′; W071°-56.23′ and 3573 masl and exhibits three terraces with several masonry walls, one with two extant niches (Fig. 7.59). The original wall likely contained two to four additional niches. The wall is oriented such that the December solstice Sun will rise behind it.

7.18 Tipon

Tipon is situated approximately 26 km to the southeast of Cusco and was an Inca city, most likely originally known as *Quispicanche*. Tipon is located at S13°-34.25′; W071°-46.99′ and 3460 masl. Zuidema (1981b, pp. 331–335) argues Tipon to have been part of a ritual pilgrimage that took place at the time of the June solstice (Sect. 3.16). The modern village of *Quispicanchis* lies along the highway in the valley below Tipon. Wright (2006) suggests that Tipon was an estate for Inca nobility. Tipon is a multifaceted complex that was found to exhibit sacred rocks, a solsticial orientation, niches, a water source, fountains, a double-jambed doorway, and terraces (Fig. 7.60).

Fig. 7.59 Wall with two niches at Kusicallanca (author's photograph)

The site includes two main sections, the lower area consisting of masonry structures, fountains fed by a mountain spring, and thirteen finely constructed terraces (Fig. 7.61). The upper area includes a complex of masonry structures with a shrine known as an *Intihuatana*. Further north is a ceremonial plaza, an aqueduct and a residential area. Above the plaza is the mountain called *Limaqawarina* (Zuidema 1981a, p. 333). Upper sections of Cusco may be seen to the northwest from both Limaqawarina and the Intihuatana.

Iglesia Raqui and Northeastern Structure

Two structures are located above the eastern flank of the terraces. One is called Iglesia Raqui (Fig. 7.65) and the other is immediately adjacent to it on its northeast side. The structures are located at S13°-34.24′; W071°-46.95′ and at 3476 masl. Iglesia Raqui has three windows and two doors facing out over the terraces and the opposite rear wall contains six windows. The northeastern structure has one door facing northeast, twelve interior niches and one window looking northwest over the terraces (Fig. 7.62).

Structures in the Intihuatana complex can be seen from Iglesia Raqui and the northeastern structure. The window in the northeastern structure faces out at 312° (Fig. 7.63) and the Intihuatana is on a 298° azimuth from the structure.

Intihuatana

Intihuatanas have been called "where the Sun is bound or measured" (Zuidema, 1981b, p. 333), but this is thought not to be an Inca term. The Tipon Intihuatana

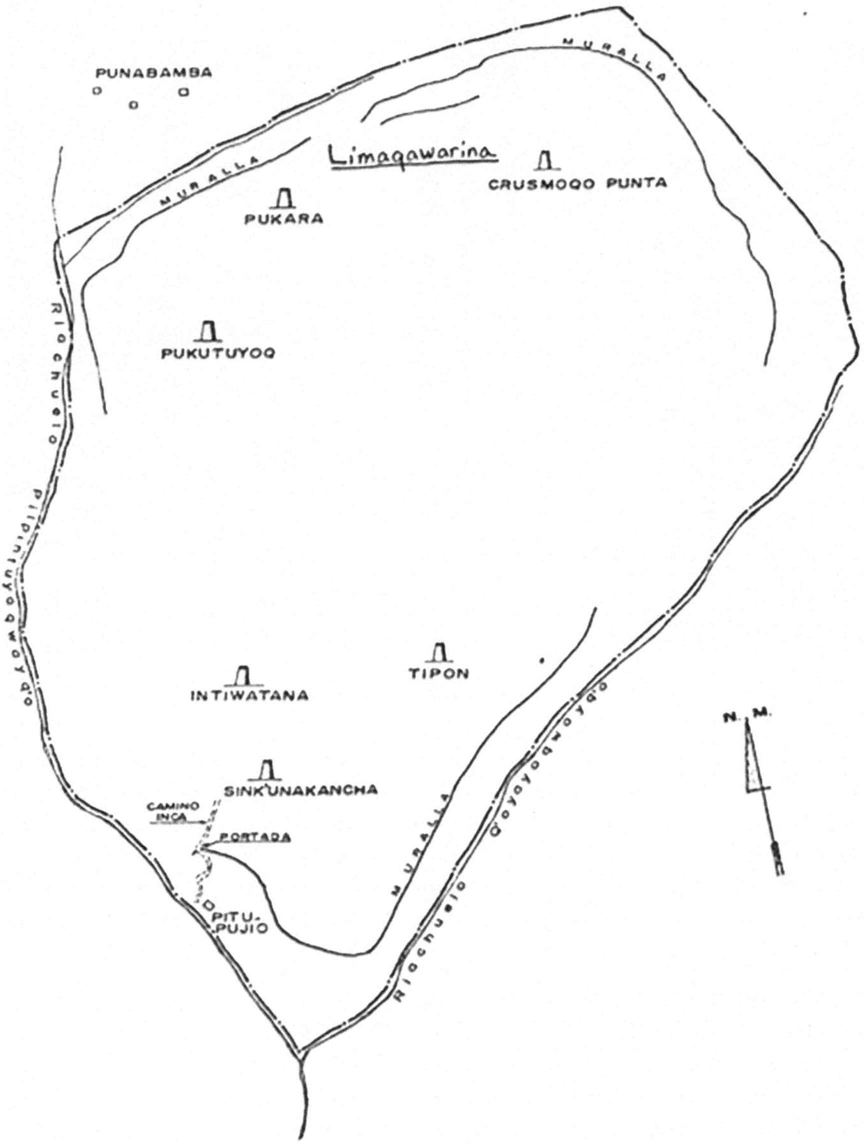

Fig. 7.60 Plan of Quispicanche, or Tipon (Fidel Ramos, from Zuidema 1981b, p. 334)

differs from the intihuatanas at Machu Picchu, the River Intihuatana, and Pisac in that those intihuatanas are all finely carved while Tipon's consists of unimproved rocks (Fig. 7.64). Similarly to the ones at Machu Picchu and Pisac, the Tipon Intihuatana is situated on high ground facilitating "access" to the Sun. It is located at S13°-34.16′; W071°-47.13′ and at 3554 masl.

Fig. 7.61 The terraces of Tipon looking to the southwest; Iglesia Raqui is to the left and the Intihuatana is to the right

Fig. 7.62 Structure immediately northeast of Iglesia Raqui (author's photograph)

Fig. 7.63 The Intihuatana as viewed from the window in the Northeastern Structure (author's photograph)

Fig. 7.64 The Intihuatana of Tipon (author's photograph)

Fig. 7.65 The eastern mesa as viewed from the Intihuatana (author's photograph)

The Intihuatana stones lie at the southern end of a complex of associated structures and overlook much of the site below. Many of the structures near the Intihuatana are crumbling and have been the subjects of reconstruction efforts. Iglesia Raqui is at a 13° angle of depression from the Intihuatana (Fig. 7.65). Tipon has a sightline aligned for the June solstice sunset that may have played a role in pilgrimage and ceremonies during the time of the festival of Inti Raymi. The Pucuy Sucanca at Quispicancha is oriented with the Chirao Sucanca at Picchu along the axis of the zenith sunrise and anti-zenith sunset (Sect. 6.7). Priests might pass through Quispicancha on their return from their pilgrimage to Vilcanota (Sect. 3.16). As with certain other Inca pilgrimage centers, there could have been separate areas for elites and non-elites to view the solstice sunset. A double-jambed doorway on the trail leading from the terraces to the Intihuatana would indicate that only elites were permitted past this point and they alone would be able to view the sunset from the Intihuatana (Fig. 7.66). Non-elites, however, could still have viewed the June solstice Sun setting over the Intihuatana from a large flat area on the site's eastern mesa near the northeastern structure next to Iglesia Raqui, a location that could accommodate a large number of observers. The orientation of the June solstice sunset axis is shown in Fig. 7.67.

Fig. 7.66 The horizon position of the June solstice sunset as viewed from Tipon's Intihuatana (author's photograph)

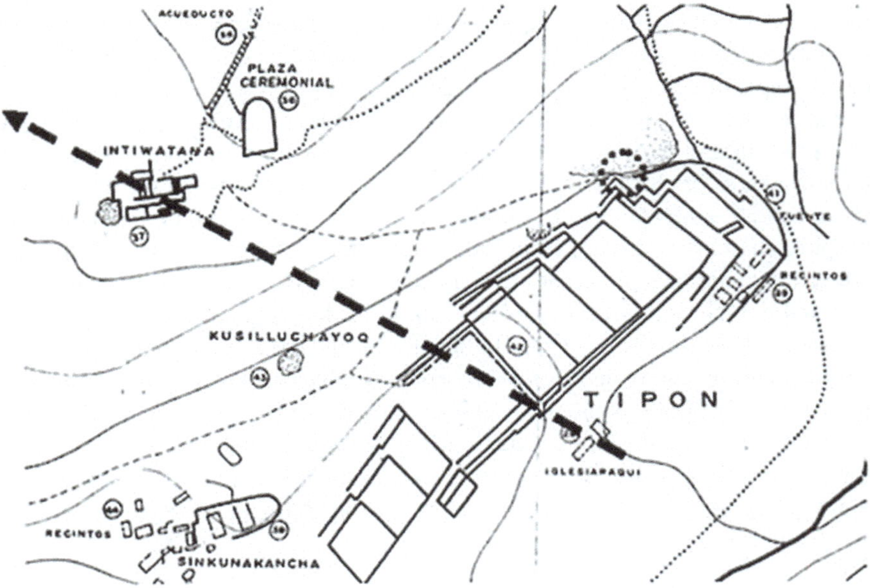

Fig. 7.67 The June solstice sunset could have been viewed over the Tipon Intihuatana from a mesa on the northeast side of Iglesia Raqui (modified from Malville 2008)

7.19 Saihuite

Saihuite, within the ancient territory of the Chanca and the modern province of Abancay, is situated some 160 km west of Cusco at S13°-32.82′; W072°-48.11′ and is at 3623 masl. Saihuite is a multifaceted complex and was found to exhibit carved rocks, light and shadow effects, a solsticial orientation, cardinal and equinoctial alignments, stairs, seats, niches, a water source, fountains, platforms, animal replica stones, and channels. The site's Principal Stone is a striking rock of andesite that has been carved extensively. The stone's upper half is completely sculpted with zoomorphic carvings as well as channels for fluids of divination.

The site is divided into hanan (upper) and hurin (lower) sectors and in addition to the Principal Stone includes the Rumihuasi ruins and an ushnu (Fig. 7.68). The Principle Stone and a structure with several rooms and a large niche rest on a raised platform in the upper sector situated on a terraced hilltop called Concacha. Beyond this platform the terrain descends rapidly and a great multi-tiered fountain that includes a staircase was constructed down the side of the hill. Located in the lower sector are several features including additional carved stones and a major platform.

Fig. 7.68 Relative positions of Saihuite features (not to scale); from left to right are the Principle Stone, the Rumihuasi Stone, the Chingana, and the Third Stone (watercolor by Jessica Gullberg)

Fig. 7.69 Saihuite's upper sector and the corners of its earthen platform on the terraced hill of Concacha. The Principal Stone is nearest the southwest corner of the platform (author's photograph)

Principal Stone

Saihuite is oriented generally southeast to northwest and the upper sector that includes the Principal Stone is at the northwest end of the site (Fig. 7.69). The features of the upper sector rest on an earthen platform. The stone is extensively sculpted and is currently protected by a wrought-iron fence (Fig. 7.70). Among the many fine carvings found on the stone are those of symbolic steps, pumas, lizards, frogs, snakes, llamas, humans, and a monkey. Carvings are in high relief, but many have been mutilated, perhaps during the period of the Spanish extirpation of idolatries (Hemming and Ranney 1982; Paternosto 1989). Van de Guchte (1990) suggests the carvings of the Principal Stone to be a figurative landscape or a map of the Inca Empire. The Principal Stone is located at S13°-32.84′; W072°-48.18′ and at 3652 masl.

A complex series of divination channels and cups begins at the top of the stone and are carved throughout (Fig. 7.71). Fluids for these ceremonies would have been fed to the channels by hand. The rock may also have been given sentience through the power of camay through water flowing in these channels across its surface.

The Principal Stone lies with the adjacent structure on the axis of the June solstice sunrise and December solstice sunset.

Upper Structure

Adjacent to the Principal Stone on the raised grassy platform is a structure with several rooms and a door with a niche facing to the southeast (Fig. 7.72). The upper sector of Saihuite is predominantly oriented to the axis of the June solstice sunrise

Fig. 7.70 The Saihuite Principal Stone (author's photograph)

Fig. 7.71 Carvings of figures, stairs and fluid channels (author's photograph)

Fig. 7.72 Elements of the upper sector of Saihuite (author's photograph)

and December solstice sunset. The long retaining walls of the earthen platform measure 065°/245° as well as do walls of the Upper Structure. The Upper Structure is located at S13°-32.83′; W072°-48.19′ and at 3657 masl.

The Upper Structure's northeast face includes a corridor and niche that are oriented to view the June solstice sunrise (Fig. 7.73). The same niche could possibly be used to watch for the rise of the Pleiades. The niche is large enough that a priest could stand within it to view the events on the distant horizon (Fig. 7.74). Additional features of the Upper Structure align on this axis as well.

Multi-Tiered Fountains

A set of 69 stairs and adjacent multi-tiered fountains connect Saihuite's upper and lower sectors (Fig. 7.75). The fountain has 22 tiers oriented to 098°/278° with a 24° slope. The lower portion of the fountain complex has six sections of two-tier fountains, while the upper portion uses three sections of three-tier fountains. There is one additional fountain in a top tier. The original source came from the terrace above and markings on the stones would indicate that water has flowed through the fountain in modern times. The top of the fountain is located at S13°-32.82′; W072°-48.14′ and 3640 masl, and the bottom is at S13°-32.82′; W072°-48.11′ and is at 3623 masl.

Fig. 7.73 Niche and corridor aligned for the June solstice sunrise (author's photograph)

Fig. 7.74 View in the direction of the June solstice sunrise from the niche (author's photograph)

Fig. 7.75 A multi-tiered fountain descends the slope leading to Saihuite's lower sector (author's photograph)

Rumihuasi Stone

The Rumihuasi Stone lies in the Rumihuasi area of Saihuite's lower sector (Figs. 7.76 and 7.77). The stone is finely carved (Fig. 7.78) on all sides and exhibits features in common with other sculptures such as the Ceremonial Rock of Machu Picchu (Fig. 9.44) and the River Intihuatana (Fig. 9.33). It exhibits shelves, seats, and symbolic steps, as well as possible divination channels on its southern side (Fig. 7.79). Fluids might have been poured into a cup on top of the rock and then flowed through a channel, first to a fork and then through either of two additional channels to small niches with drains. Paternosto (1989) writes that the stone is still used for marriage ceremonies, the bride and groom each drinking water from the two channels. No apparent celestial orientations were found in the Rumihuasi Stone and it now is split in two. The Rumihhuasi Stone is located at S13°-32.77′; W072°-47.97′ and at 3584 masl.

Chingana

The platform of Chingana ("where one gets lost") in the lower sector is cardinally oriented (Figs. 7.80 and 7.81), with the long walls measuring east-west and the shorter ones nearly north-south. The rear wall of the platform extends for 38 meters at 090°/270°. The front wall measures 091°/271°. Both side walls are 21 meters long at 008°/188°. The staircase includes 16 steps that are eight meters wide and face 003°.

Fig. 7.76 The lower sector of Saihuite (author's photograph)

Fig. 7.77 Area surrounding the Rumihuasi Stone (author's photograph)

Fig. 7.78 The Rumihuasi Stone of Saihuite (author's photograph)

Fig. 7.79 Ceremonial channels in the Rumihuasi Stone (author's photograph)

Fig. 7.80 *Chingana* in Saihuite's lower sector. A depression can be seen in the center (author's photograph)

Fig. 7.81 Chingana as viewed from front left (author's photograph)

Fig. 7.82 The Third Stone's front side (author's photograph)

Platforms, as ushnus, were used for ceremony and sacrifice, but the specific function of this site remains unclear. It is a raised platform with stairs that has at its center a one square meter depression that may have served to drain fluids – two of the three elements typical of an ushnu. Chingana's size would also make it suitable as a place to gather for viewing ceremonies held on Ushnu Pampa immediately to its north. On the north side of Ushnu Pampa are the remains of a considerably deteriorated platform called Rauraq'asa. Chingana is located at S13°-32.88′; W072°-47.89′ and at 3565 masl.

Third Stone

Paternosto (1989) says the Third Stone (Figs. 7.82 and 7.83) was first discussed in 1972 by Hébert-Stevens. Hemming and Ranney (1982, p. 176) refer to it as "a pumalike stone at Saihuite" but only in the caption of one of their photos. Paternosto (1989, p. 135) continues that "this Third Stone is the best instrument for watching the movement of the sun between the solstices, as the shadow projected by the stepped quarter-disc – the 'gnomon' – demonstrates." Paternosto refers to following the shadow of the central arm as if it were a sundial depicting the extremes at solstices and without shadow at the equinoxes. Staller (2008, p. 298) says that the stone's carvings are oriented in such a way that it can serve observation of the annual cycle of solar movement. As such it acts as a gnomon or pillar, the last of the three prescribed elements for an ushnu at Saihuite.

Fig. 7.83 The Third Stone's rear side (author's photograph)

The Third Stone is finely carved on all sides. The axis of its central arm is 057°/237°, or a 33° difference from the 090°/270° equinoctial axis. The arm with a hole suggests potential use as an intihuatana, but the true purpose of the Third Stone remains enigmatic. The stone lies at the opposite extreme of Saihuite from the Principal Stone. The Third Stone is located at S13°-32.84′; W072°-47.79′ and is at 3545 masl.

7.20 Summary

The region surrounding Cusco exhibits many fine examples of Inca astronomy encoded into its huacas and other structures. The "Eyes of the Puma" at Kenko Grande and the solar orientations within that huaca's cavern readily demonstrate the Incas' fascination with the Sun and their propensity for utilizing their astronomical knowledge when constructing important shrines. Nowhere better is this shown than within the caves of Lacco where the altars are illuminated by the Sun or the Moon at specific times of the year.

Huaca 44 with its gnomons aligned for solar horizon events is especially interesting. The Incas designed multiple meaning and purpose into most everything they carved and this huaca is certainly no exception. The astronomical orientation of the gnomons and the sculpted seats are unmistakable and were prominent features of this site. The full significance of this huaca, however, is as of yet unknown and remains to be determined.

At Tipon the alignment for the viewing of the June solstice sunset by elites and potentially non-elites as well is intriguing as this becomes part of an emerging trend found to be common with other pilgrimage sites. In Saihuite the most prominent of solar axes, that between the horizon points of the June solstice sunrise and December solstice sunset, dominates the upper sector of the complex. The Incas were very aware of the travel through the sky of their Sun-god and honored him with ceremonies and observances guided by alignments that they designed into many of their carvings and structures.

Chapter 8
Orientations in the Sacred Valley

Fig. 8.1 Boy by ceremonial zigzagged channel at Chinchero (watercolor by Jessica Gullberg)

Pachacuti celebrated his victory over the Cuyos by establishing an estate and palace at Pisac. Later he constructed Ollantaytambo after conquering the Tambos and apparently built Machu Picchu to commemorate further conquests in the Vilcabamba (Niles 1999; Rowe 1990). Pachacuti's son, Topa Inca, established a royal estate at Chinchero and his grandson, Huayna Capac, did so in Urubamba. Machu Picchu is explored in Chap. 9, while the rest are among the sites presented in this chapter (Fig. 8.2).

© Springer Nature Switzerland AG 2020

S. R. Gullberg, *Astronomy of the Inca Empire*, Historical & Cultural Astronomy,
https://doi.org/10.1007/978-3-030-48366-1_8

Fig. 8.2 Sites profiled in Chap. 8 (watercolor map by Jessica Gullberg). Legend: 8.1 Chinchero*. 8.2 Pisac*. 8.3 Q'espiwanka*. 8.4 Cerro Pumahuachana. 8.5 Cerro Unoraqui*. 8.6 Maras. 8.7 Moray*. 8.8 Choquequilla*. 8.9 Ollantaytambo*. *Astronomical orientation identified

8.1 Chinchero

One of the first tasks before each new Inca was the establishment of his royal residence (Niles 1999). Topa Inca, the son of Pachacuti, claimed the Chinchero valley as the site for his estate and soon set about construction of his palace, courtyard, support buildings, and agricultural terraces. Also at the site are several intricately carved rock huacas. The style of architecture and design suggests a view of nature similar to that of Pachacuti, in which natural rock and landscape features were included in structural forms. Chinchero is located at S13°-23.40′; W072°-02.78′ and is at 3790 masl. It is a multifaceted complex that was found to exhibit light and shadow effects, solsticial orientations, cardinal and equinoctial orientations, stairs, seats, niches, a water source, basins, caves, altars, animal replica stones, channels, and terraces.

The sites of Pachacuti are steeped with solsticial alignments, but his son, Topa Inca, may have embraced a different philosophy. The many walls and terraces of the Chinchero complex are oriented cardinally, north-south and east-west. This in itself may represent a significant use of astronomical alignment as the cardinal direction of south could first have been determined by the shadow plot of a vertical gnomon. North, east, and west would then follow geometrically. Alternately, east and west could have been traced on the day of an equinox.

Chinchero was the only site in the Sacred Valley found to exhibit extensive cardinal orientations, but Morris and Thompson (1985, p. 60) tell of similar alignments at Huánaco Pampa. They describe an ushnu and gateways as being aligned approximately on an axis of 088° and mention this as being close to the direction of the equinox sunrise. Remains of a canal were found near the center of the gateways with a similar east – west orientation. Specific Inca interest in the equinoxes is subject to debate, however, as the Spanish chronicles make little mention of equinoctial usage.

There are two major carved rocks, the first to the south of the plaza, Titikaka, has two carved stairways, one of which leads upward to the top of the rock through a tunnel-like passage with an approximately north-south axis. On the top there are a series of cut rectangular trays similar to those of Kenko Grande and Lacco. To the southwest is a second large carved stone, Chinkana, containing an elaborate stairway, enclosures, altars, and carved trays. At its lower end is a flowing stream beneath carved teeth. A triangular basin opens approximately to sunset on the December solstice. Above and to the south are carved stones known as Mesakaka and Kondorkaka. While the majority of the features of Chinchero emphasize cardinal directions, solsticial alignments are not totally absent. The two primary carved rock huacas, Titikaka and Chinkana, lie on the axis of the June solstice sunrise and December solstice sunset.

The primary entry to Chinchero opens onto a town plaza that serves as a market with a modern building situated on the edge of its lower western half. A Catholic chapel was built upon an earlier Inca structure on the eastern upper extreme. The upper and lower terraces of the plaza are separated by a retaining wall containing three doorways and twelve large niches (Fig. 8.3). This wall is aligned north and south along its length.

North of the town plaza and chapel are the remains of three rectangular structures oriented east-west, and beyond this is the great plaza that Gasparini and Margolies (1980) tell us is now called *Capallanpampa* (Fig. 8.4). The orientation is shown in Fig. 8.5. Further to the east is an expansive system of agricultural terraces retained by walls of fine masonry. These walls are also oriented to the east and west and the carved rock huacas of Chinchero surround them.

Cardinal and Solsticial Alignments

The palace, terraces, and structure walls near the plazas tend to have prominent east-west orientations (Fig. 8.6) with other walls oriented north-south. The huacas of Mesakaka and Kondorkaka are aligned south of the Chinkana stone.

It could have been possible to determine east-west at times of the equinoxes when a gnomon cast its shadow in these cardinal directions. Finding true south was perhaps even more definitive by tracking the gnomon's shadow over a period of time and recording the endpoint of its midday position each day, as described in Sect. 5.4. A parabolic curve is the ultimate result, and while the Incas likely did not

Fig. 8.3 Town plaza retaining wall at Chinchero (author's photograph)

Fig. 8.4 East-west structural remains and the plaza of Capallanpampa (author's photograph)

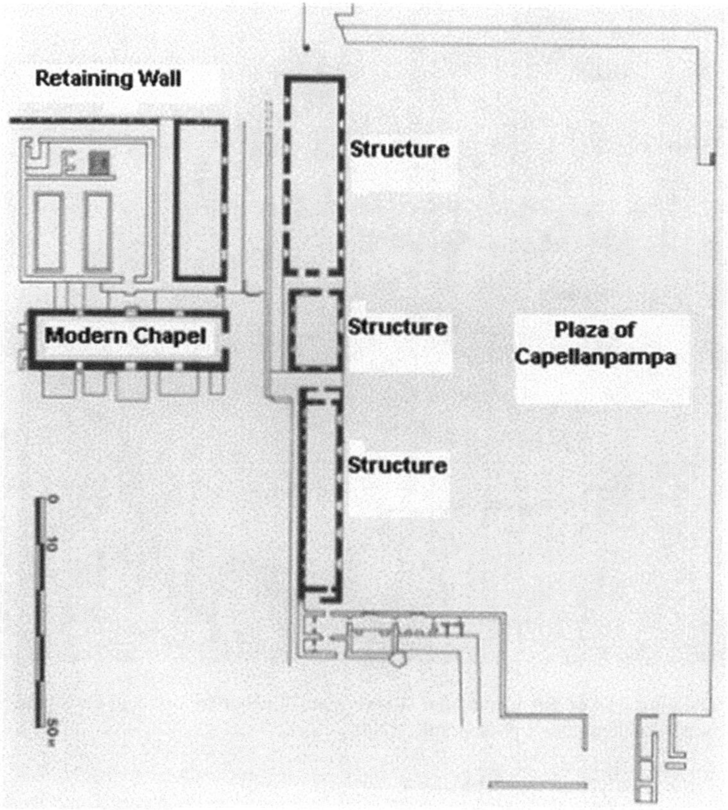

Fig. 8.5 Plan of Chinchero (modified from Gasparini and Margolies 1980)

recognize parabolas as such, they certainly could recognize the point in the curve closest to the gnomon. A line drawn from the gnomon through this point is aligned directly with true south. The Incas then could find east-west as being perpendicular with this south-north axis, but there is no historical reference of them using such techniques.

Solsticial Alignments

The two primary carved rock huacas of Chinchero, Titikaka and Chinkana, are aligned on the axis of the June solstice sunrise and December solstice sunset (Fig. 8.7). Both of these huacas are very large and were carved in situ.

Fig. 8.6 The long wall of this rectangular structure in Chinchero is oriented east-west, while the shorter is north-south (author's photograph)

Fig. 8.7 Titikaka and the direction of December solstice sunset as viewed from the top of Chinkana (author's photograph)

Fig. 8.8 The zigzagged channel at Chinchero (author's photograph)

Zigzagged Channel

Located near the main entrance to the town plaza is a carved zigzagged offeratory/divinatory channel (Fig. 8.8). It is reminiscent of the one found atop Kenko Grande, but this one could have been transported to this spot from another location. Liquids would have been poured into the cup on the upper side by the rock and then observed as they flowed to the opposite end.

The First Rock East of the Chapel

Once beyond the modern chapel heading to the east, the first carved rock exhibits a giant seat facing back to the southwest (Fig. 8.9). The seat is oriented to 253°, near enough to easily view the December solstice sunset, but the outcrop appears to exhibit no specific astronomically related alignments. The First Rock is located at S13-23.40°; W072-02.78° and at 3790 masl.

Fig. 8.9 The First Rock east of the chapel at Chinchero (author's photograph)

Kondorkaka

Continuing east and at the end of the upper terraces is a huaca called *Kondorkaka* (Fig. 8.10). Kondorkaka is significantly eroded, but still exhibits carved seats facing to the north, northeast, and to Nevado Chicon. Several seats were sculpted on this side of the outcrop, with one of them oriented to the mountain. Near the base are some carvings of animals. I found no astronomical orientations, but Kondorkaka lies on a north-south line with Mesakaka and Chinkana. Kondorkaka is located at S13-23.35°; W072-02.61° and 3773 masl.

Mesakaka

North of Kondorkaka and heading downslope is *Mesakaka* (Fig. 8.11), a carved rock with many square tables or trays. In keeping with the predominant theme of Chinchero, these squares also are oriented generally east-west and north-south. The view aligns perfectly to overlook Chinchero's central valley and sunsets on the western horizon (Fig. 8.12). Immediately on Mesakaka's northern side is a smaller rock with a single seat. Mesakaka is located at S13-23-29°; W072-02.59° and is at 3744 masl.

Fig. 8.10 Kondorkaka (author's photograph)

Fig. 8.11 Mesakaka (author's photograph)

Fig. 8.12 The western view from Mesakaka (author's photograph)

Chinkana

Chinkana is an outcropping of limestone and is one of the two principle huacas of Chinchero (Fig. 8.13). Chinkana lies low on the eastern end of Chinchero's central valley and has been extensively carved on all sides with such as seats, stairs, shelves, and niches. A stream was routed to pass beneath the base of the rock on its western side, likely to energize the huaca through camay (Fig. 8.14). A niche by the creek faces 350° and a notch on the horizon is inclined by +12°. Chinkana faces Titikaka in the direction of the December solstice sunset.

Fig. 8.13 The western face of Chinkana (author's photograph)

Fig. 8.14 A stream runs along the base of Chinkana to the right of a niche (author's photograph)

Fig. 8.15 A carving resembling teeth on Chinkana's northern face (author's photograph)

On Chinkana's lower northern face is carved a large but shallow niche with a row of what resembles teeth across its top (Fig. 8.15). The niche is carved east-west and faces north. A functional staircase descends from the top of the huaca and winds from north around to the west as it descends to a balcony carved into the side of the stone. The balcony looks up the valley and faces 278°, giving it a splendid view of Chinchero sunsets.

A triangular basin exists on a separate stone immediately northwest of Chinkana (Fig. 8.16). This basin faces 238° and is oriented upward 5°, generally toward Titikaka and the December solstice sunset. A seat carved on top of Chinkana at its northern edge faces 020° and Nevado Chicon. The northern edge of the base of Chinkana is located at S13°-23.27′; W072°-02.58′ and is at 3724 masl.

Titikaka

Approximately 360 meters west of Chinkana and north of the great plaza of Capallanpampa is Chinchero's largest carved outcrop, *Titikaka* (Fig. 8.17). Titikaka displays many carvings including two prominent stairways, one external, and the other within a break in the center of the rock leading to its top (Fig. 8.18). The central stairway exhibits figurative carving and is oriented for a solar effect during June solstice.

Fig. 8.16 The triangular basin (author's photograph)

Fig. 8.17 Titikaka (author's photograph)

Fig. 8.18 The central stairway of Titikaka (author's photograph)

The upper surface of the stone is extensively carved and displays several examples of seats or trays and animals such as a condor and a snake. Titikaka and Chinkana both lie on the axis of the June solstice sunrise and December solstice sunset. There is a deep hole, perhaps to hold a pole for a textile banner, carved on the top of Titikaka that may have helped to mark this alignment. The bearing to Chinkana and the triangular basin, as measured from this hole, aligns with the June solstice sunrise (Fig. 8.19).

At the base of the rock, lower on the western side, are more carvings, a large crevasse and an opening to a shallow cave (Fig. 8.20). The crevasse looks out at 278° and the cave opens to 254°. The cave will receive light at the time of, but is not perfectly aligned for, the December solstice sunset. A large niche was carved near the mouth of the cave and looks out at 293°. The lower base of Titikaka is located at S13°-23.35′; W072°-02.80′ and is at 3753 masl.

8.2 Pisac

Seventeen kilometers northeast of Cusco and high above the Vilcanota River lay the ruins of Pisac, with its impressive mountainside terracing (Fig. 8.21), its many examples of fine masonry, and a carved shrine known as the Intihuatana. Niles (1999) states that Pisac was established by Pachacuti in commemoration of his

Fig. 8.19 Chinkana aligns with the June solstice sunrise as seen from Titikaka and measured by Kim Malville (author's photograph)

Fig. 8.20 Crevasse and cave opening in Titkaka's lower western face (author's photograph)

Fig. 8.21 The terraces of Pisac (author's photograph)

defeat of the Cuyos. Pisac is a multifaceted complex and was found to exhibit carved rocks, light and shadow effects, solsticial orientations, an equinoctial orientation, niches, a water source, fountains, and platforms. Pisac's Intihuatana is located at S13°-24.86′; W071°-50.64′ and is at 3390 masl.

Intihuatanas, or "hitching places of the Sun," (Hemming and Ranney 1982) are also found at Machu Picchu (Fig. 9.7), the Urubamba River Intihuatana near Machu Picchu (Fig. 9.33), and Tipon (Fig. 7.64) and possibly were places of solar worship. In Pisac the intihuatana is a large, partially carved rock in the temple group that is enclosed by a semicircular masonry wall adjoining a straight masonry wall in the form of the letter "D" (Figs. 8.22 and 8.23). It displays a circular stone carving on its flat upper surface within the walled enclosure. Squire (1878) wrote that he was informed by the Governor of Pisac that this gnomon had once been clad with a bronze sheath, but there are no substantiations for this claim. From the gnomon a nearby peak can be seen in the direction of the June solstice sunrise, although evidence of such usage remains to be seen. The primary rock extends beneath and beyond the wall of the structure where a second gnomon was sculpted. Dearborn and Schreiber (1986) describe similarities in design between this intihuatana and the Torreon of Machu Picchu (Fig. 9.14).

Immediately to the east of the Intihuatana is another enclosed rock which appears to be an observing platform or usnu. It opens on the southeastern horizon toward the 114° direction of December solstice sunrise. Pisac has no elaborately carved huacas, but its fine masonry and the water channels designed to bring it life indicate the site's cultural importance.

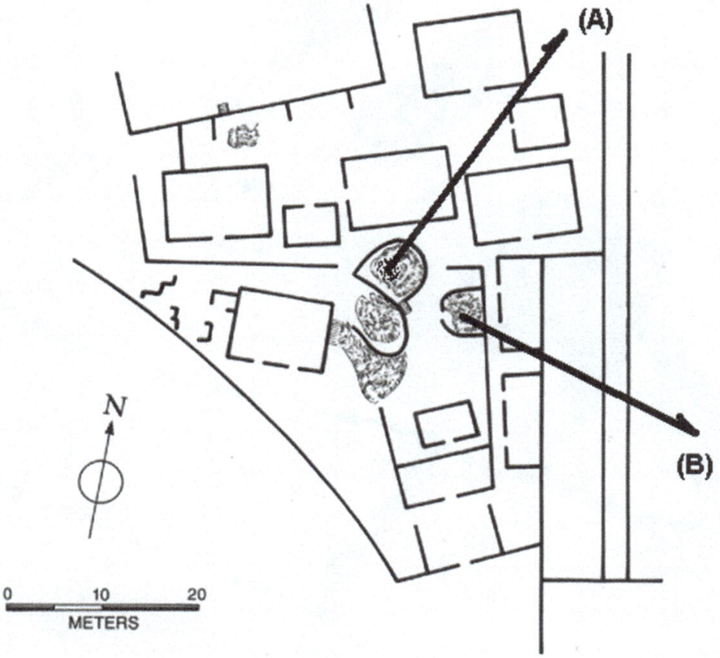

Fig. 8.22 Plan of Pisac. (**a**) is the direction of the June solstice sunrise from the Intihuatana. (**b**) Is the direction of the December solstice sunrise from the platform (modified from Hemming and Ranney 1982)

Fig. 8.23 The Intihuatana of Pisac (author's photograph)

Fig. 8.24 The chakana near Pisac's Intihuatana (author's photograph)

On the western side of the carved rock and below the secondary gnomon is a carving of the upper half of a chakana, or Inca cross (Figs. 8.23 and 3.36). The sculpture is oriented 090°/270° and according to a local informant the lower half of the cross is completed by a properly aligned shadow on the day of an equinox.

While the Intihuatana could be considered as a temple of the Sun, to its west is a structure presently called the "Temple of the Moon" (Fig. 8.25). Hemming and Ranney (1982) refer to this as the "chamber with two doors." This building's walls were constructed of fine masonry and the interior sides contain 17 niches. Two doorways were built into the front of the structure and open to 211°. The temple could be illuminated by moonlight, but there are no obvious orientations.

8.3 Q'espiwanka

Q'espiwanka is a multifaceted complex and was found to exhibit a sacred rock, solsticial orientations, niches, a water source, pillars, double and triple-jambed doorways and structures. It was a palace of Huayna Capac and is located at S13°-18.05′; W072°-06.71′ and is at 2934 masl. Niles (1999) made a detailed study of Q'espiwanka's physical designs as they appear in the archaeological record. Huayna Capac began construction of Q'espiwanka some time prior to 1499 CE, three to four years after the death of his mother. At that time he was involved in a

Fig. 8.25 Pisac's Temple of the Moon (author's photograph)

military campaign against the Chachapoyas, and the design of his royal palaces was apparently left to one of his half-brothers, Lord Sinchi Roca.

The estate spanned the Yucay Valley, combining some 20 separate properties of maize fields and included both the valley floor and agricultural terraces (Farrington 1995). Farrington identifies three sectors: (A) an agricultural zone to the north, (B) the structural zone, and (C) the zone south of the palace where the remains of a reservoir has been found. The majority of the palace grounds consisted of a great plaza (in sector B), at the center of which was a large granite boulder (Fig. 8.26) and, as suggested by Niles (1999), a platform which has now been replaced by a modern chapel (Fig. 8.27). The eastern wall of the plaza was an impressive sight with twin double-jambed gate houses, a large ramp, and a triple-jambed entranceway. As Niles (1999) describes it, the architectural design of the palace's eastern face was designed to impress whomever approached. Upon entering the main gateway a visitor would first see the huge plaza, perhaps empty except for the granite boulder and its adjacent platform. The platform was in the geometric center of the plaza, along the centerline of the triple-jambed entry portal, and it seems likely that the precise placement of the palace and the plaza was established by the location of the boulder. It is thought that the name Q'espiwanka was taken for this sacred boulder, with "q'espi" the polishing of stone surfaces and "wanka" a culturally important large stone (Farrington 2017). Niles (1999) suggests that a channel was designed to carry water from an area in the north to the boulder and its platform.

Fig. 8.26 Central granite boulder at Q'espiwanka (author's photograph)

Solar Pillars

On top of the Cerro Sayhua ridge (modern nomenclature) northeast of Q'espiwanka there are two pillars long known to the inhabitants of the nearby modern village of Urubamba, and easily visible from below (Fig. 8.28). There is ethnohistorical description by Spanish chroniclers of similar pillars surrounding the city of Cusco (Sect. 6.7), but none of those structures remain. If Q'espiwanka's pillars were similar in design and intent to those that have been reported on the horizons of Cusco, then they would be two of the only extant examples. Although known to the local community and identified by Bauer and Dearborn (1995, p. 69) as "useful examples of what Inca solar pillars may have looked like," the astronomical functions of these pillars had not been fully ascertained until more recently. Zawaski (2007) observed that the Urubamba pillars marked June solstice sunrise and it was later determined that the large granite boulder (Fig. 8.26) in the courtyard of Q'espiwanka is the properly aligned point for viewing this phenomenon. Stone-lined channels in the courtyard of Q'espiwanka could have surrounded the boulder with water, and today a modern channel carries water past its flank. The boulder appears to be the primary point in the area for observing June solstice sunrise with the horizon pillars. Figure 8.29 shows the Sun rising along the eastern pillar on 8 June 2008, nearing the end of a northward journey that culminates at the pillars during the time of the solstice.

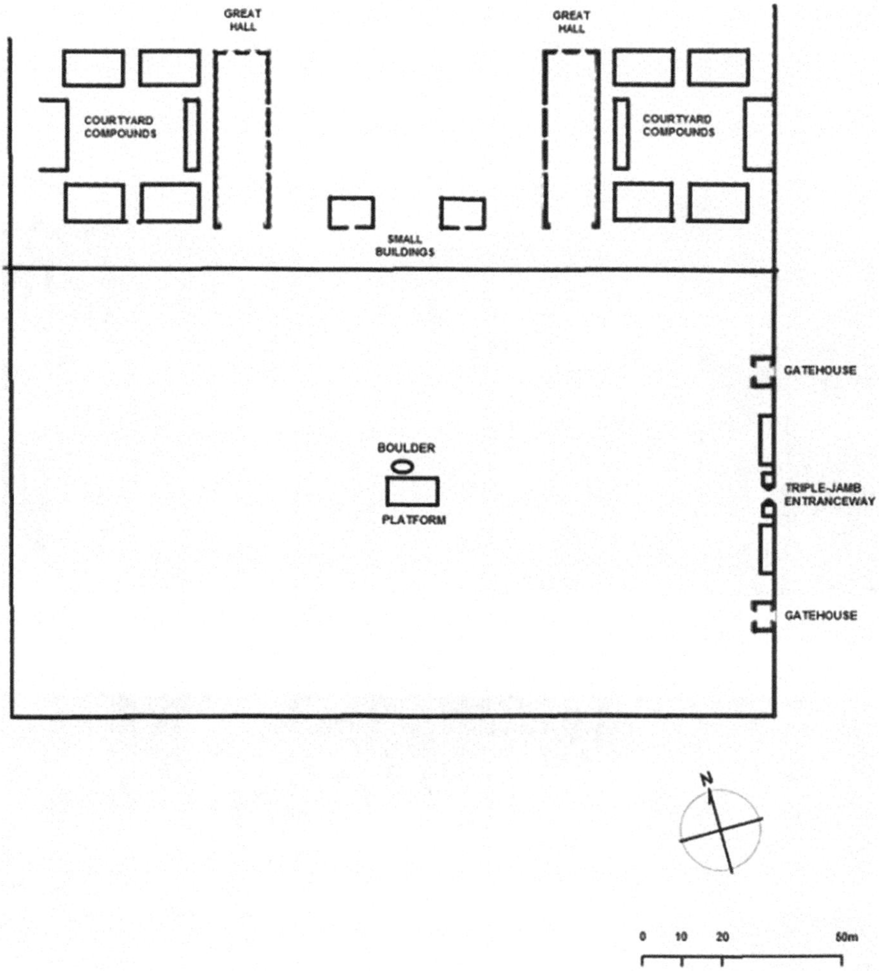

Fig. 8.27 Plan of Sector B of Q'espiwanka (modified from Niles 1999)

The granite boulder sits near the center of the ancient plaza and immediately adjacent to a modern chapel which Niles (1999, p. 173) suggests was the location of Q'espiwanka's platform. The boulder is 5.15 meters long, 3.73 meters wide and 2.47 meters high. It is located at S13°-18.05′; W072°-06.71′ and at 2934 masl.

Natural features on the ridgeline alone might have been sufficient for priests to determine the time of the approaching solstice. Construction of the towers underscores the importance that the Inca placed on the June solar event and the significant role that this site may have played in their ritual. The pillars could have been designed to be a visual element of the Inti Raymi ceremonies held below.

As viewed from the granite boulder, the azimuth of the midpoint of the pillars is 057° at an inclination of 22° above level. The pillars are 35 meters apart on either

Fig. 8.28 The pillars on Cerro Sayhua (author's photograph)

Fig. 8.29 Sunrise over the right Cerro Sayhua Ridge pillar as viewed from Q'espiwanka's white boulder in June prior to the solstice; the sunrise each day will continue to move toward the left pillar (visible to the left of the Sun) until the time of the solstice (author's photograph)

Fig. 8.30 The eastern pillar above Q'espiwanka (author's photograph)

end of a level area and are constructed of sandstone (Fig. 8.30). They are aligned with each other on an azimuth of 100° at an elevation of approximately 3860 meters. The eastern pillar has a height of 4.3 meters and a base 1.5 meters by 3.3 meters; the base of the partially restored western pillar is similar.

Viewing From the South Wall of the Palace

Q'espiwanka's courtyard may not have been an area intended for public ceremonies because the palace's eastern wall contains a massive triple-jambed doorway (Fig. 8.31) surrounded by two double-jambed gatehouses. Inca doorways with multiple jambs typically marked entry into a space used only by elites. The courtyard may thus have been similar in ceremonial use to that of the Sanctuary of the Island of the Sun at Lake Titicaca in that non-elites were restricted from entry (Dearborn, Seddon, and Bauer 1998).

June solstice sunrise would have also been viewable from outside the southern wall of the palace. The wall is oriented 110°/290° and Niles (1999) suggests that

Fig. 8.31 Remains of the triple-jambed entranceway in Q'espiwanka's eastern wall (author's photograph)

there were 40 double-jambed niches along its 190-meter length (Fig. 8.32). The wall faced an artificial lake and large granite boulders. This area may have been used for public viewing as a place where pilgrims and other non-elites were allowed to observe the solstice sunrise between the pillars in a manner similar to ceremonies held on the Island of the Sun (Dearborn, Seddon, and Bauer 1998). As with the platform alignments at the Island of the Sun, the public viewing area for the June solstice sunrise outside of Q'espiwanka might have been designed to feature the Sun rising over Huayna Capac, the ruling Inca and son of the Sun, perhaps as he stood upon the platform proposed by Niles (Fig. 8.33). A potential viewing location outside the southern wall is at S13°-18.08′; W072°-06.78′ at 2922 masl.

A potential third site for the June solstice solar ceremonies could have been a 40-meter-long terraced platform built between and around the pillars. Both ends of the sightline connecting the palace and the pillars could have been huacas. As part of the celebration a group of celebrants could climb the 950 meters from the valley to the platform where they could then place offerings, make sacrifices, and celebrate the passage of the Sun as it traveled across the sky from dawn to dusk.

The question arises as to how the Urubamba pillars escaped the Catholic extirpation of idolatries that destroyed the pillars of Cusco. One possibility is that they were relatively modest features on the high horizon that escaped significant attention from below.

Fig. 8.32 Q'espiwanka's southern wall (author's photograph)

Fig. 8.33 The rise of the June solstice Sun over the pillars and southern wall (author's photograph)

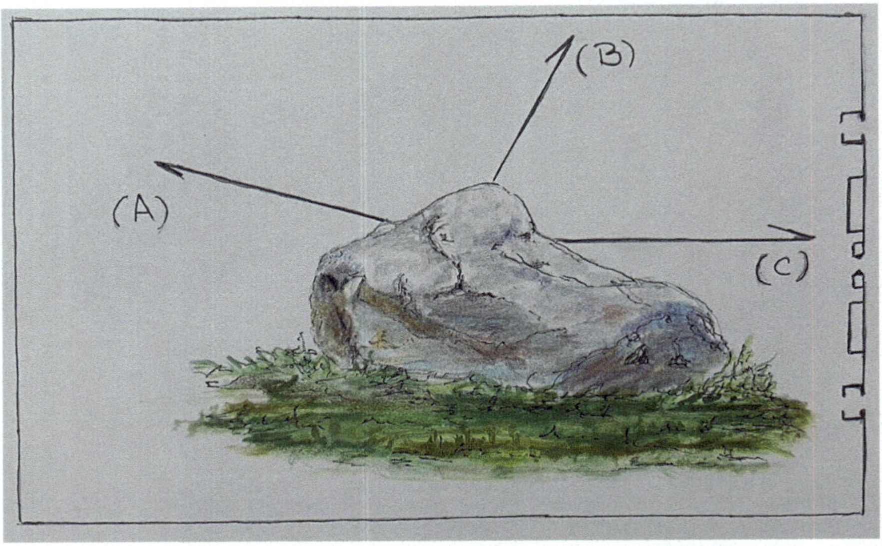

Fig. 8.34 General directions from Q'espiwanka's boulder for the June solstice sunrise (**b**), the December solstice sunrise (**c**), and the June solstice sunset (**a**) (watercolor by Jessica Gullberg from Niles 1999)

Other Alignments

The granite boulder in Q'espiwanka's courtyard may also have been a viewing point for other solar orientations with respect to natural features (Figs. 8.34 and 8.35). The boulder is situated to observe the December solstice sunrise on the horizon near the 12 km-distant peak of Cerro Unoraqui (Figs. 8.36 and 8.37). Cerro Unoraqui is discussed further in Sect. 8.5.

The 294° June solstice Sun sets over a nearby peak when witnessed from Q'espiwanka's granite boulder (Fig. 8.38). Natural features on the mountain could have made the point of sunset predictable but are subtle enough to remain inconclusive. There is no direct evidence of Inca interest in this observation.

The horizons around Q'espiwanka were evaluated with potential natural features for marking the solar horizon positions of the December solstice sunset, the equinox sunrise, equinox sunset, the zenith sunrise and the anti-zenith sunset. In each case a subtle feature of the landscape on the horizon could have identified the event, but none of the features stood out enough to support such usage.

8.4 Cerro Pumahuachana

Cerro Pumahuachana looms above the village of Yucay and the palace of Sayri Tupac. The ridge is oriented approximately 025°/205° (Fig. 8.39) and harbors the remains of two stone wall segments. The center of Cerro Pumahuachana is located

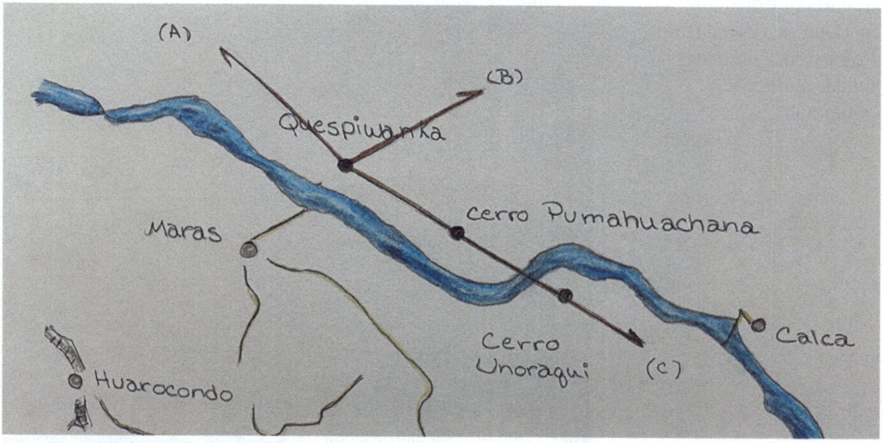

Fig. 8.35 The alignment of Cerro Unoraqui as viewed across Cerro Pumahuachana from Q'espiwanka in the direction (**c**) of the December solstice sunrise (**b**) is the direction of the June solstice sunrise and (**a**) the June solstice sunset (watercolor by Jessica Gullberg)

Fig. 8.36 Cerro Unoraqui as viewed from Q'espiwanka. Cerro Pumahuachana is the ridgeline in the foreground (author's photograph)

Fig. 8.37 December solstice sunrise in the direction of Cerro Unoraqui as viewed from Q'espiwanka's granite boulder (author's photograph)

Fig. 8.38 June solstice sunset from Q'espiwanka's granite boulder (author's photograph)

Fig. 8.39 Cerro Pumahuachana ridge (author's photograph)

approximately at S13°-19.03′; W072°-03.92′ and is at 3575 masl. It was found to exhibit structures and is situated along the solsticial sightline between Q'espiwanka and Cerro Unoraqui.

A modern chapel is located on a crag above the ridge's northern end. Extant remains of walls are at ground level and stretch for over 200 meters along the top of the ridgeline (Fig. 8.40). Local villagers relate that rocks from the walls were taken to the site of the chapel and were used in the construction of its base. While this chapel is modern, it may have been built on an ancient site of Inca interest (Fig. 8.41). As of yet, such historical evidence has not been found.

The chapel is clearly visible through the southern double-jambed gate house at Q'espiwanka (Fig. 8.42) and would be as well through the triple-jambed entrance, if the view were not presently blocked by trees. Relatively little remains of the northern gate house. The alignment of the palace's eastern wall and its structures is oriented in the direction of the crag with the chapel.

When Cerro Unoraqui is viewed from the granite boulder at Q'espiwanka the line of sight crosses Cerro Pumahuachana near an 85-meter gap between the two major segments of wall remains (Fig. 8.34). The crag and chapel are several hundred meters to the left of the line of sight to Unoraqui.

Fig. 8.40 Wall remnants on Cerro Pumahuachana (author's photograph)

8.5 Cerro Unoraqui

Cerro Unoraqui is a 4377 masl peak on the southern side of the Urubamba River between Yucay and Calca (Fig. 8.43). Its pillars are located at the summit at S13-20.36°; W072-00.44°and it exhibits solsticial sightlines and cardinal alignments.

Fig. 8.41 Chapel on a crag of Cerro Pumahuachana (author's photograph)

Fig. 8.42 The chapel as viewed through the southern double-jambed gate house (author's photograph)

Fig. 8.43 Cerro Unoraqui from Cerro Pumahuachana (author's photograph)

As viewed from the granite boulder at Q'espiwanka, Cerro Unoraqui is aligned across the Pumahuachana ridge for the December solstice sunrise, thus providing a natural feature that could have been used to mark the position of the solstice (Fig. 8.34).

Atop the summit there are three rock pillars local villagers claim originated in the time of the Incas, but this is unsubstantiated (Fig. 8.44). The pillars are encircled by a low stone wall and are aligned on a north-south axis with respect to one another. Such north and south alignments may suggest possibilities relative to Inca interest in the Southern Cross and certain Inca dark constellations within the Milky Way. Equinox sunrises and sunsets could also be viewed across the pillars.

The wall around the pillars is oval in shape with its longer axis 10.70 meters and the shorter 7.80 meters. The walls of the oval average 0.80 meters in height. From south to north each pillar is on slightly higher ground. The south pillar has a base of 1.60 meters and is 1.73 meters high. The mid pillar's base is 1.80 meters and its height is 1.70 meters. The north pillar is 2.30 meters wide and 2.20 meters high. The north and mid pillars are 1.10 meters apart and there are 0.70 meters between the mid and south pillars. The north and south pillars are each 1.60 meters from the oval wall along its long axis. On the short axis the mid pillar is 4.20 meters from the east wall and 3.60 meters from the west. The pillars are aligned north and south.

Fig. 8.44 Pillars on top of Cerro Unoraqui (author's photograph)

Dominating Cerro Unoraqui's northern horizon is the Cordillera Urubamba, including the 5530 meter Nevado Chicon and the 5818 meter Nevado Ccolque Cruz (Fig. 8.45).

8.6 Maras

Since the time of the Incas and before, salt from a subterranean stream has been mined through evaporation in terraced ponds near the village of Maras. There are thousands of individual salt ponds owned by members of the community (Fig. 8.46). While some have tried to put forth myth of astronomy at the site, no significant evidence has been found for such. The water source functions to provide the salt to the ponds. The site is located approximately at S13°-17.92′; W072°-09.40′ and is at 3380 masl.

8.7 Moray

Topa Inca created terraced basins in three natural sinkholes near his estate of Chinchero at a site called Moray (Fig. 8.47). Moray is located approximately at S13°-19.82′; W072°-11.75′ and at 3385 masl.

Fig. 8.45 Cerro Unoraqui pillars aligned north-south (author's photograph)

Fig. 8.46 Salt ponds of Maras (author's photograph)

Fig. 8.47 Terraces and basin at Moray (author's photograph)

Ken Wright (2011), of the Wright Paleohydrological Institute, concluded that these terraces could not have been used for agricultural purposes because water would have to have flowed uphill to irrigate the northern terraces. This led Wright and his associates to believe the function of the site to be mainly ceremonial.

Moray includes several modified boulders including one referred to as The Iconic Rock (Christie 2013). Astronomical usage has not been identified.

Malville (2015) suggests that a possible ceremony could have occurred at the time of zenith Suns. On these two days each year the terraces briefly cast no shadows, thus sunlight would fall straight to the Earth and the Pachamama within (Fig. 8.48).

A June solstice sunrise effect is enabled by nine terraces, which are oriented for the rising Sun and the related shadows (Figs. 8.49 and 8.50). The smallest of Moray's muyus, muyu D, is aligned as a potential point for viewing this sunrise.

Fig. 8.48 Sunlight would enter the terraced basin directly from above at the time of the zenith Sun (author's photograph)

Fig. 8.49 Terraces for the June solstice sunrise effect (author's photograph)

Fig. 8.50 Moray's June solstice sunrise effect (from Malville 2015)

8.8 Choquequilla

Above the Rio Huarocondo, 5 km southeast of Ollantaytambo and 14 km west of Urubamba, are the ruins of Choquequilla. Choquequilla is located at S13°-17.53′; W79°-13.93′ and is at 3627 masl. It is a complex that was found to exhibit a carved rock, a solsticial alignment, niches, a water source, a cave, and terraces.

The remote huaca lies within the mouth of a cave opening oriented to the approximate direction of the December solstice sunrise. The intricately carved shrine faces inward toward the cave, away from the horizon, and is flanked to the south by a wall constructed with two rows of four double-jambed niches, emphasizing the site's significance. The roof of the cave is formed by two relatively flat stone faces that form an inverted "V" (Fig. 8.51). Light from the December solstice Sun as it rises above the opposing horizon brightly illuminates the cave and huaca.

The cave is situated on a mountainside above agricultural terraces that have fallen into disuse. A central staircase ascends the terraces and at the top a trail proceeds north to the cave. A small masonry structure with a door and windows is situated immediately to the cave's north. The carved rock of black granite (Fig. 8.52) is said to be among the finest examples in existence and exhibits great symmetry and exquisite carving (Paternosto 1989). The sculpting greatly

Fig. 8.51 Carlos Aranibar at the mouth of the cave of Choquequilla (author's photograph)

Fig. 8.52 The carved black granite rock and wall with niches at Choquequilla (author's photograph)

resembles that of the Baño de la Ñusta at Ollantaytambo (Fig. 8.63), but the Choquequilla rock was damaged, possibly by looters.

Paternosto (1989, p. 89) calls this "the cave of *Choqequilla*, the Golden Moon," and Van de Guchte (1990, p. 191) calls it the "Moon Temple" of Choquequilla. Both imply lunar ceremonies. The cave opens to the December solstice sunrise, while the carved stone is slightly offset at 130° and the horizon is elevated by 32°. The rise of the solstice Sun illuminates the cave brightly (Fig. 8.53). The light of a rising full Moon could also create a dramatic effect.

8.9 Ollantaytambo

Ollantaytambo is located at S13°-15.95′; W072°-16.03′ and is at 2818 masl. It is about 90 km from Cusco in the Urubamba River valley. Niles (1999) tells us that Pachacuti developed the area after defeating the Tambos and that both Huayna Capac and Manco Inca made subsequent improvements. This was the site of one of the few great victories of the Incas over the Spaniards when Manco Inca had the plain below flooded, thus miring the conquistadors' invading horses (Hemming and Ranney 1982). Ollantaytambo is a multifaceted complex that was found to exhibit carved rocks, light and shadow effects, solsticial orientations, zenith alignments, seats, a water source, fountains, gnomons, structures, and terraces.

Terraces

The most striking feature when first approaching Ollantaytambo is a magnificent set of 17 stone terraces that ascend the hillside (Fig. 8.54). The extensive Terraces of Pumatillis exhibit alignments that average approximately 114°, thus facing out to the rise of the December solstice Sun and, in the opposite direction, facing toward and framing nicely the 294° June solstice sunset. A staircase begins right of center and continues along the left side of the terraces. The terraces would have provided an excellent platform for a large group of people to view the sunrise during the coming of age festivities at the time of Capac Raymi at December solstice.

Temple of the Sun

What is known as Ollantaytambo's Temple of the Sun (Fig. 8.55) was extensively damaged by the Spanish in their purge of indigenous religion, however a foundation and a wall of six monoliths survive (Hemming and Ranney 1982). The wall is located to the left of the top of the Terraces of Pumatillis and faces the Pinkuylluna

Fig. 8.53 The December solstice sunrise brightly illuminates Choquequilla's cave (author's photograph)

mountain in a direction that approaches alignment with the 063° rise of the June solstice Sun, but is not close enough to be considered as such an orientation (Fig. 8.56). A specific purpose for viewing a solstice sunrise from this vantage point remains unclear. Hanzalova, Klokocnik, and Kostelecký (2014) instead propose the observation point to be Ollantaytambo's Pyramid of Pacaritanpu.

Fig. 8.54 The Terraces of Pumatillis at Ollantaytambo (author's photograph)

Fig. 8.55 Ruins known as the Temple of the Sun (courtesy of Carlos Aranibar)

Fig. 8.56 June solstice sunrise over the Pinkuylluna mountain and the Huaca of Chimpaccahua (author's photograph)

Pinkuylluna Mountain

The Pinkuylluna mountain lies opposite Ollantaytambo to the northeast and aligns with the June solstice sunrise as viewed from the Pyramid of Pacaritanpu (Fig. 8.57). The mountain exhibits two structures and a face glaring ominously from its side (Fig. 8.58). It is said that the face is that of Tunupa, the messenger of the god Wiracochan, turned to stone (Salazar and Salazar 2014).

Fig. 8.57 Pinkuylluna mountain (author's photograph)

Fig. 8.58 **The face of Tunupa on** Pinkuylluna mountain (author's photograph)

Fig. 8.59 The illuminated Pyramid of Pacaritanpu corner at June solstice is circled; the shaft of light passes nearby the Temple of the Sun and the face of Viracocha (Salazar and Salazar from Hanzalova et al. 2015, p. 42)

> In the Andean cosmology these conversions to stone, which is to say the passage from one form of life to another, does not imply a definite separation from this world, but rather it is assumed that from their new state those beings continue to contribute to the dynamic process of life (Salazar and Salazar 2014, p. 30)

Pyramid of Pacaritanpu

Salazar and Salazar (2014) show orientations in Ollantaytambo for the June solstice sunrise (Fig. 8.59) and the December solstice sunrise (Fig. 8.60) at the Pyramid of Pacaritanpu. They state that Pacaritanpu means "Lodging of the Dawn or Palace of Windows" (Salazar and Salazar 2014, p. 97). Sunlight is cast upon a corner of the pyramid at the time of the June solstice and then again across its valley face at December solstice. Salazar and Salazar continue that the pyramid was designed to be seen from high above at the huaca of Intipuncu. Hanzalova, Klokocnik and Kostelecký (2015) have also performed research that supports these solar orientations at the solstices. Salazar and Salazar (2014) continue that Inca cosmology relates that at dawn on the June solstice light passes between the adjacent mountains to illuminate "the place where tradition said the Incas first entered the world as gods" (p. 100).

Fig. 8.60 December solstice sunrise across the valley with the pyramid (Salazar and Salazar from Hanzalova et al. 2015, p. 44)

Incamisana

Paternosto (1989) says that while the Temple of the Sun was the primary site for ceremony in Ollantaytambo's hanan, or upper, sector, the Incamisana was its ceremonial counterpart in the lower hurin sector. Salazar and Salazar (2014) call this "…Inticcahuarina, or the place from where one sees and understands the sun" (p. 115). Horizontal gnomons project distinct shadow effects at two times of solar significance. A local informant tells that at the time of the equinoxes a face is projected toward the lower left in shadow by the gnomons and rock. The face is said to be that of a man playing a quena, or flute. On the December solstice at local noon the shadow of one of the gnomons is said to reach down and "insert" itself to fill a carved triangular notch in the base below. A second notch may have existed, but if so is now broken (Figs. 8.61 and 8.62). At the time of the zenith Sun the shadows of the three lower gnomons are said to touch each of the three lower steps.

Bañjo de la Ñusta

The Bañjo de la Ñusta, or "Bath of the Princess," is a striking fountain set in Ollantaytambo's lower urban sector (Fig. 8.63). Ollantaytambo contains many water channels and fountains and the Bañjo de la Ñusta is noteworthy for its sculptural similarity with the 5 km-distant black granite rock at Choquequilla (Fig. 8.52).

Fig. 8.61 The horizontal gnomons of the Incamisana (author's photograph)

8.10 Summary

Numerous examples of astronomical orientations exist in the Sacred Valley. Chinchero's many masonry walls exhibit cardinal orientations of north and south that may have been determined by astronomical means. The site's two major huacas also take advantage of an alignment for the June solstice sunrise and December solstice sunset. Significant interest in this sunset is evidenced by a balcony carved on Chinkana that faces it, by a nearby triangular basin that is appropriately aligned, and by the shelves or trays of Mesakaka that are oriented in this direction as well.

Pisac displays a carved stone called the Intihuatana that has a circular gnomon on its top and is aligned with a nearby peak for the June solstice sunrise. Additionally, an adjacent platform opens to the rise of the Sun on the December solstice.

The finely carved huaca of Choquequilla is located in the mouth of a cave that is oriented with the December solstice sunrise and is brilliantly illuminated at that time. Ollantaytambo is a great site of astronomy. Its outstanding terraces of Pumatillis align for the December solstice sunrise and June solstice sunset, while the Pyramid of Pacaritanpu is oriented for June solstice sunrise over the nearby Pinkuylluna mountain and December solstice sunrise through the valley. The gnomons of Ollantaytambo's Incamisana exhibit calendrical shadow effects during solar events.

Fig. 8.62 The notched shadows (Salazar and Salazar from Malville 2015)

Fig. 8.63 The Bañjo de la Ñusta (author's photograph)

The most significant astronomical orientation discovered in the Sacred Valley, however, is that of the solar pillars above Q'espiwanka. These structures are extant examples of a type of horizon astronomy reported to have existed extensively around Cusco. The pillars are aligned so that the Sun will rise there on the morning of the June solstice as viewed from a platform that was adjacent to the palace's granite boulder. It was also possible to view the December solstice sunrise over the distant Cerro Unoraqui from this same location. Non-elites may have viewed the June sunrise from a separate location outside the palace's southern wall, as entry to the main plaza could have been restricted to elites only by virtue of the symbolic double and triple-jambed entries.

Chapter 9
Orientations in and Surrounding Machu Picchu

Fig. 9.1 Machu Picchu Ceremonial Rock (Pencil by Jessica Gullberg)

Machu Picchu has been variously described as a citadel (Bingham 1930), a lost city (Bingham 1948), a royal estate (Rowe 1990), and a sacred center (Reinhard 2002). Machu Picchu was likely recognized by the Incas as a place of power because of its many granite outcrops and caves and the cardinality of mountains such as Veronica,

© Springer Nature Switzerland AG 2020

S. R. Gullberg, *Astronomy of the Inca Empire*, Historical & Cultural Astronomy,
https://doi.org/10.1007/978-3-030-48366-1_9

Machu Picchu, Huayna Picchu, Pumasillo, and the snow peak of Salcantay. Salazar and Salazar (2014) state that this is a sacred place and that the mountains Huayna Picchu and Machu Picchu form a "divine pair." They continue that considerations such as this played a part in the decision to build here. They also state that in Quechua "Machu Pichiu" means "Old Bird" and that the general ground plan of the sacred site resembles the form of a condor when viewed from above, such as from the top of Huayna Picchu.

Salazar (2004) notes the large number of shrines at Machu Picchu, similar to the huacas that surround Cusco, and acknowledges that more such shrines exist at Machu Picchu than are found at other royals estates such as Pisac, Chinchero, Huamanmarcha, and Callachaca.

Machu Picchu may also have been the focal point of a regional pilgrimage center that included Llactapata (Fig. 9.2). Pilgrimage is discussed in Sect. 3.16.

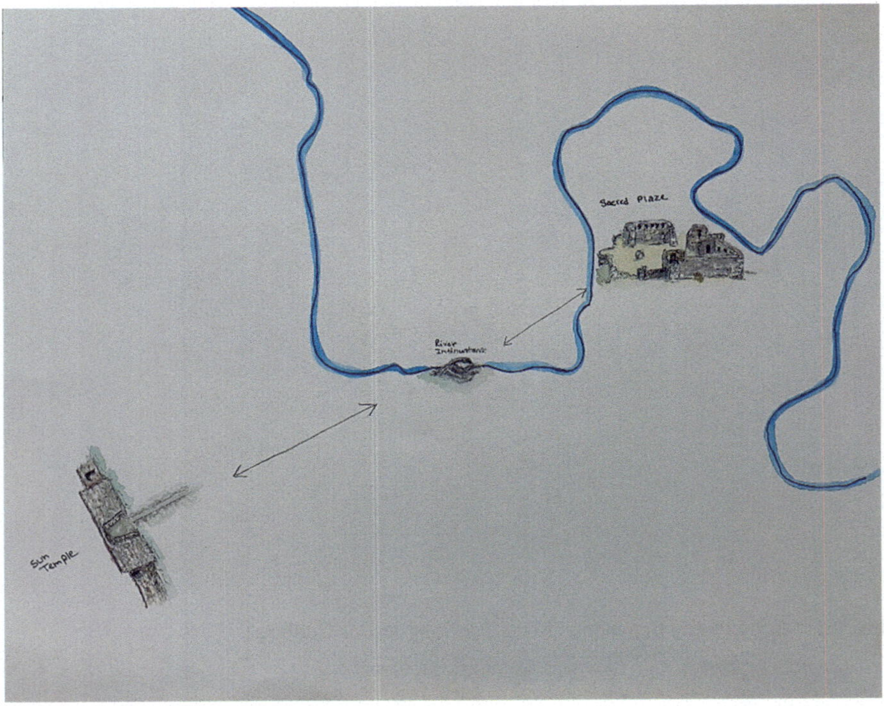

Fig. 9.2 Relative positions of the Llactapata Sun Temple, the River Intihuatana, and the Machu Picchu Sacred Plaza on the axis of the June solstice sunrise and December solstice sunset (Watercolor by Jessica Gullberg)

9.1 History and Design

The Inca trail was the primary means of access to Machu Picchu from Cusco. Pilgrims could have arrived on the horizon through the Sungate and then would have proceeded past a check post to the mountain shrine. The main section of Machu Picchu could have served as an elite pilgrimage area with access through its principal doorway. Others may have remained outside in the vicinity of the Ceremonial Rock (Fig. 9.37) or possibly continued along the trail across the drawbridge (Fig. 9.3; Fig. 3.35) and on to Llactapata.

Niles (2004) notes that the series of royal estates in the Vilcanota-Urubamba Valley were a string of named places similar to the sequential organization of huacas along ceques in the Cusco valley, all of which could be recorded on quipus. Pachacuti's assimilation of the Cuyos was commemorated by the establishment of an estate at Pisac. When the Tambos were conquered, he is said to have built his estate at Ollantaytambo. Finally, he may have constructed Machu Picchu when he obtained control of the Vilcabamba (Niles 1999, p. 133). Pachacuti's son, Topa Inca, later established his estate at Chinchero, and his grandson, Huayna Capac, followed with his palace, Q'espiwanka.

A major function of these royal estates was to produce food and some, such as at Topa Inca's palace, exhibited extensive terraces and others, such as Pachacuti's

Fig. 9.3 Author and research assistant wife, Jessica, by the drawbridge and trail on the west side of Machu Picchu (Author's photograph)

estate at Ollantaytambo and Huayna Capac's at Q'espiwanka, had rich and well-irrigated river bottom land. Topa Inca's properties also included salt production at the salt flats near Maras. Niles (2004, p. 60) argues that the estates supported significant ritual activities required by petitions to certain gods and regularly occurring events of the sacred calendar and "...in addition to the inherent sacredness of a particular location on an estate, estates likely had provisions for royal worship." It thus is likely that ritual activities were major features of Machu Picchu.

The handiwork of Pachacuti can be seen in the elaborately carved huacas of the Cusco valley and the precisely fitted andesite blocks of the Coricancha. The stone work of Pisac is similarly skillfully executed, but there seem to be fewer examples of symbolic iconic motifs except for water channels and the intihuatana enclosure. The next royal estate to be established by Pachacuti, Ollayatambo, has an abundance of iconic features such as the massive stones of the "Sun Temple," elaborate fountains and water channels, sections of stairways in the cliffs, as well as a horizontal gnomon. But it is in his estate of Machu Picchu that his apparent penchant for improving on the "handiwork of the creator" is fully evident. Perhaps it was because Machu Picchu offers much in terms of granite outcrops, boulders, and caves that Pachacuti committed great effort to such creative activity.

Machu Picchu may have been vacated as the empire fell during the Spanish conquest (Rowe 1990; Wright and Valencia 2000). Opinions vary as to whether the departure took place during or prior to the conquest. In either case the abandonment is likely a significant reason that the conquistadors never found the site. There is speculation that the abandonment might have been due to a smallpox epidemic that swept the empire prior to the European invasion. Another reasonable alternative is that Machu Picchu was Pachacuti's estate. In accordance with Inca custom it was not inherited by his son, Topa Inca, and instead would have become the responsibility of Pachacuti's panaca. Topa Inca built his own country estate and thus Machu Picchu may have simply been deserted for reasons of disuse. Machu Picchu remains a previously undisturbed example of Inca culture and, as such, a valuable location for research of Andean astronomy.

Architectural Plan and Engineering

The site for Machu Picchu lies some 500 meters above the Urubamba River between two reverse faults that have given it the drop-down area between them. Fracturing at the faults created building material for Inca masons and facilitated the spring necessary for the site's water supply (Wright and Valencia 2000). When selecting this location, Pachacuti would have been impressed by the magnificent views of several sacred mountains.

The estate appears to have been carefully planned and was likely built with mitmae workers. Like Cusco, Machu Picchu was divided into hanan and hurin sectors (Gasparini & Margolies 1980). Principal temples and palaces were constructed in the hanan sector.

Extensive agricultural terraces were built above and below Machu Picchu. These were constructed for erosion control, aesthetic appeal, and crop production. The terraces, however, were not capable of producing enough to feed the site's population, thus a great amount of food products would have been brought in from below (Wright and Valencia 2000).

Ceques, Axes, and Sightlines

The number of huacas in and around Machu Picchu, while fewer, is reminiscent of the concentration of huacas emanating from the Coricancha of Cusco where 41 ceques originate. Juan Polo de Ondegardo (1965[1571], p. 67) wrote "In each village the organization was the same; the district was crosscut by ceques and lines connecting shrines or various consecrations and all the things which seem notable: wells and springs and stones, hollows and valleys and summits which they call apachetas." Ceques have many interpretations, one of which is that of visual alignments. It is possible that ceques may have existed at Machu Picchu and one such potential orientation involves Machu Picchu, Llactapata, and the River Intihuatana.

The River Intihuatana lies below the intersection of two major axes, one of which is shown in Fig. 9.2. This is a solar axis formed between the horizon points of the June solstice sunrise and December solstice sunset, which also includes Llactapata's Sun Temple and Machu Picchu's Sacred Plaza. The second (not shown in Fig. 9.2) is an equinoctial east-west axis proceeding from Llactapata's Overlook Temple directly across the River Intihuatana and on to Cerro Machu Picchu, but this alignment is uncertain as it would cross the solsticial one and ceques are generally thought not to cross one another. If that were true, however, this equinoctially oriented line still could have been an axis while not at the same time a ceque. It remains possible that the River Intihuatana's location stems directly from such relationships, though, and that it may have been part of a ceque connecting Machu Picchu with Llactapata. It is also a possibility that the precise geographic location of the River Intihuatana at the junction of such axes was intentional and that this may be an example of the power and status of this huaca.

Gary Ziegler (personal communication 2008) feels that a petroglyph (Fig. 9.4) located north of Machu Picchu's main plaza serves as a map and that the essentially straight lines etched upon it represent the ceque system of Machu Picchu. The carvings have suffered much erosion but are distinctly reminiscent of the strings of a quipu, or of a ceque system such the one found in Cusco. Ziegler says that two of the petroglyph's sixteen lines are directionally oriented for the June solstice sunrise/December solstice sunset axis, while others of those that remain point to mountains and other alignments. This concept is unproven and remains a subject of Ziegler's continuing research.

Fig. 9.5 displays the view from Machu Picchu's Sacred Plaza down to the River Intihuatana as part of a potential ceque. Figure 9.6 shows the reverse sightline from the River Intihuatana up to Machu Picchu.

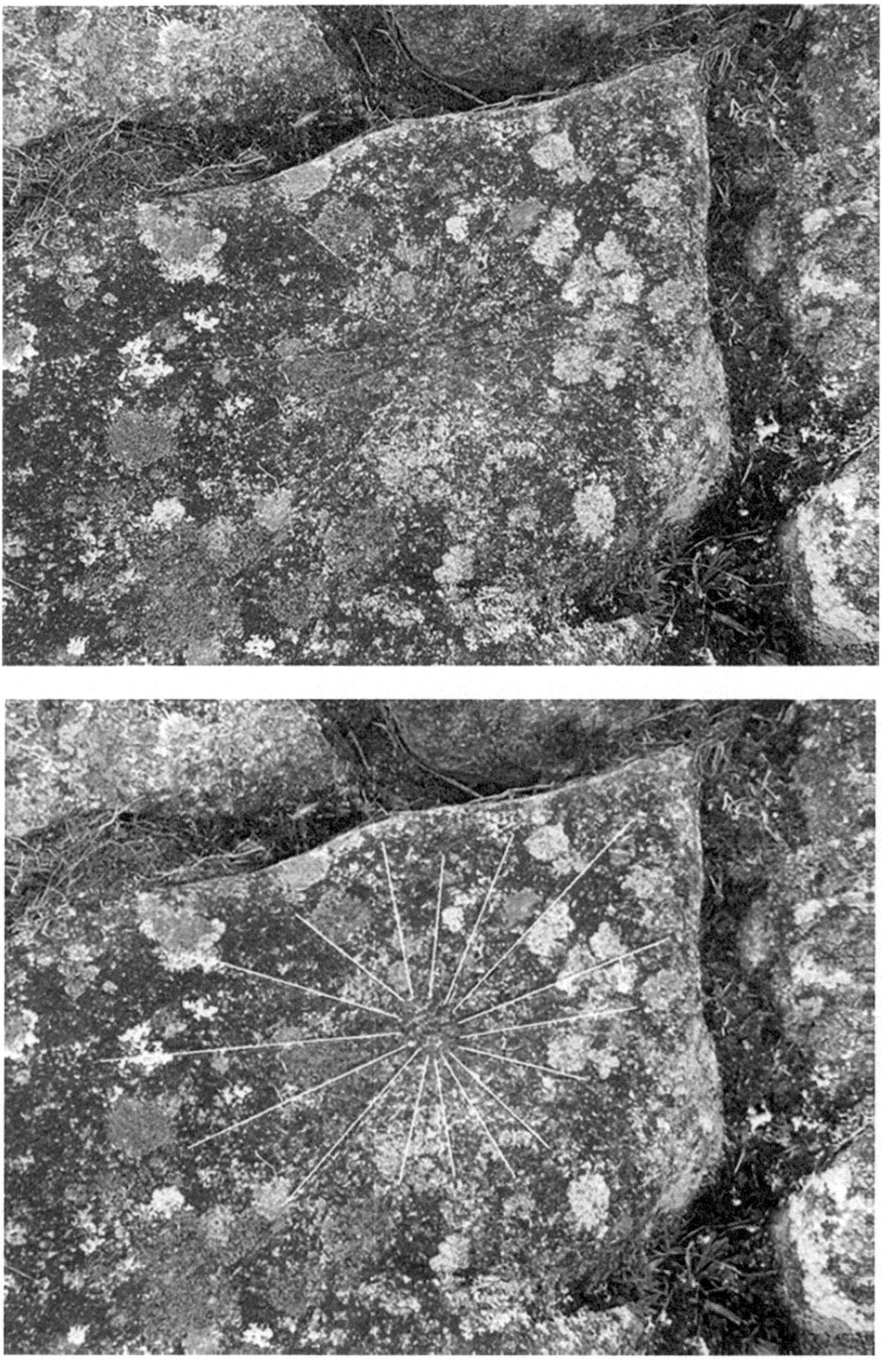

Fig. 9.4 Faint lines carved on the petroglyph at Machu Picchu in the upper image are enhanced in the lower one (Author's photograph)

Fig. 9.5 The River Intihuatana as seen from Machu Picchu's Sacred Plaza; the southwestern terraces from which a trail leads down to the river are visible in the foreground. Also visible is the bridge of PeruRail (Author's photograph)

Fig. 9.6 Machu Picchu as viewed from the River Intihuatana (Author's photograph)

9.2 Machu Picchu Site

Machu Picchu is located at S13°-09.79′; W072°-32.74′ and is at 2468 masl. It is a multifaceted complex that exhibits carved rocks, light and shadow effects, solsticial orientations, an equinoctial orientation, a zenith alignment, an anti-zenith orientation, a Pleiades orientation, stairs, seats, niches, a water source, fountains, basins, caves, altars, platforms, double-jambed doorways, animal replica stones, horizon replica stones, structures, and terraces.

Intihuatana.

The Intihuatana is a focal point of Machu Picchu that has a commanding view of the surrounding area and the mountains and valleys beyond (Fig. 9.7). It was carved in situ and exhibits a vertical column on top of a sculpted platform. Intihuatanas are discussed in Sect. 3.5.

Intihuatanas are also found by the Urubamba River (Fig. 9.31), at Pisac (Fig. 8.23), and at Tipon (Fig. 7.64). In Pisac the intihuatana is a large, partially carved rock that is enclosed by a semicircular masonry wall that adjoins a straight masonry

Fig. 9.7 Looking across the Intihuatana of Machu Picchu and a gorge to the site of the Sun Temple of Llactapata (Author's photograph)

wall. There is a vertical stone on its flat upper surface. The primary rock extends beneath and beyond the wall of the structure where a second gnomon is located. The Intihuatana of Tipon exhibits a different style in that the in situ rock remains unimproved although a platform has been built around it. The Pisac intihuatana shares structural similarities with Machu Picchu's Torreon.

Evidence has not been found to indicate that Machu Picchu's intihuatana served as an astronomical observing instrument, but its unobstructed views would have made it valuable for various observations. On an expedition with Dearborn, as viewed from the intihuatana, Bernard Bell observed two pillar-like objects on San Miguel. These were not substantiated, but such observations are possible from such a vantage point (Dearborn and White 1982).

The Intihuatana was built upon the highest point in Machu Picchu's urban sector and is positioned well for visual alignments with Llactapata and the River Intihuatana. The view from Machu Picchu's Intihuatana to Llactapata is inclined by 2°. The River Intihuatana is lower by 20°. The June solstice sunrise is easily viewed from the Intihuatana in the opposite direction on a horizon measuring 13° up on a 064.5° bearing. A characteristic common among most of the rocks identified as intihuatana is that their sizes are small enough that they could be finely carved on all four sides. Machu Picchu's Intihuatana is located at S13–09.79°; W072–32.74° and is at 2468 masl.

Sacred Plaza

Below Machu Picchu's Intihuatana to the south is the Sacred Plaza (Fig. 9.8). The Sacred Plaza contains the Principle Temple and the Temple of Three Windows. The Temple of Three Windows faces toward Llactapata to the southwest and the setting Sun at December solstice. The axis of the plaza and the walls of the temples are aligned with the axis connecting June solstice sunrise and December solstice sunset (Figs. 9.9 and 9.10). On the Sacred Plaza's southwest side is a semicircular platform open to December solstice sunset that overlooks the Urubamba valley with views of the River Intihuatana and the Llactapata Sun Temple.

The Sacred Plaza may be part of an intentional ceque that includes the River Intihuatana and the Llactapata Sun Temple; all three are aligned with the axis of the June solstice sunrise and December solstice sunset. A possible Inca June solstice ceremony has been suggested that may have taken advantage of the alignment and elevation of Llactapata's Sun Temple as viewed from the Sacred Plaza. On the morning of the June solstice the Sun's rays first strike Llactapata before illuminating Machu Picchu. Hypothetically, golden medallions could have cast a brilliant reflection of the Llactapata dawn's first light back to onlookers at the Sacred Plaza. This is discussed further in Sect. 9.3.

The Llactapata Sun Temple bears 245° from the Sacred Plaza and is inclined 2.5° higher. The River Huaca is situated 20° below the Sacred Plaza on a 224° azimuth. The Sacred Plaza is located at S13–09.82°; W072–32.73° and is at 2460 masl.

Fig. 9.8 Machu Picchu's Sacred Plaza orientations (From author's photograph)

Torreon/Royal Mausoleum/Temple of the Sun

The Torreon and Royal Mausoleum was one of the major shrines of Machu Picchu (Fig. 9.11). Very fine masonry crowns the top where it has been carefully fitted into the rock with a wall that includes a window open to June solstice sunrise and the heliacal rise of the Pleiades. There are two other windows on this level, one with unusual holes that may have been used for attaching religious objects. There also is an arrow stone that points toward Mt. Salcantay. The rock surrounded by the walled enclosure is illuminated through the June solstice-oriented window on that day at sunrise (Fig. 9.12). A ledge cut into the top of the rock nearly bisects the early rays of the solstice Sun (Dearborn and White 1989). This boulder may have played a role in ceremonies conducted by priests on this upper level of the structure. Below the Torreon is the Royal Mausoleum, a cave that contains a set of symbolic stairs, niches, and stonework crafted with the existing boulder (Fig. 9.13). The cave gives the impression that a passageway to the underworld has been revealed. The outer rock appears to have been peeled away, exposing finely carved interior spaces that may portray the underworld from which ancestors of the Incas are said to have emerged (Salazar 2004). The carved, stepped stone may represent a shamanic stairway of ascent, and/or the route of ascent of the ancestors from the world below. The close proximity of the Torreon to the major fountains of Machu Picchu is consistent with the importance of water in the symbolic meaning of the rock.

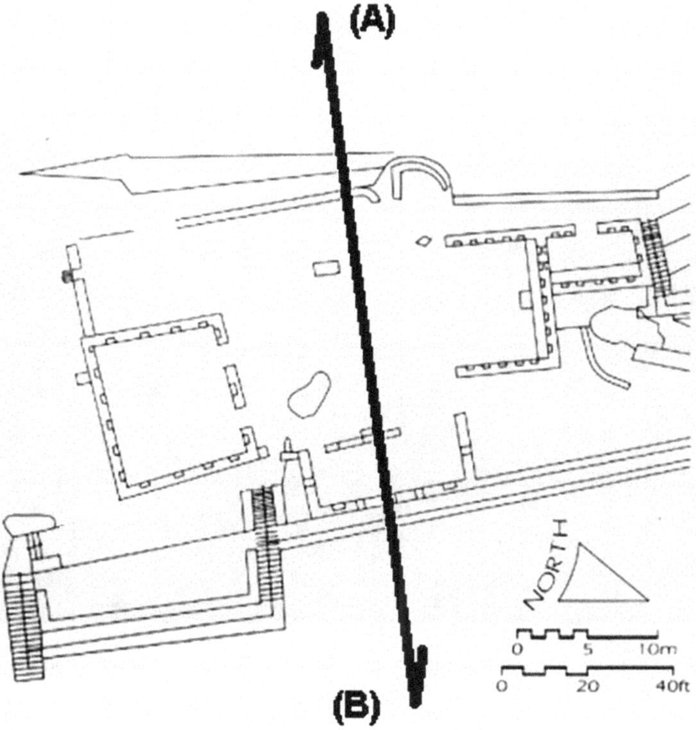

Fig. 9.9 Plan of the Sacred Plaza; the general axis of the (**a**) December solstice sunset and the (**b**) June solstice sunrise is depicted (modified from Wright and Valencia 2001)

The Torreon may have served to observe zenith passage as well as the June solstice. Research at Pisac's Intihuatana (Fig. 8.23) revealed a structure with similar characteristics (Dearborn and Schreiber 1986). Comparisons have also been made between the Torreon and the Coricancha with regard to structure and Pleaides observations. The respective horizon points of the June solstice sunrise and the heliacal rise of the Pleiades are both visible through the Torreon's window. Dearborn and White found the window to be more closely aligned with that of the Pleaides, while the ledge on the rock is oriented for the solstice (Fig. 9.14).

The site of the Torreon is a promontory with an eastern view. The semi-circular structure includes two windows – one oriented to the northeast and the other to the southeast (Dearborn and White 1982; Hanzalova, Klokocnik, and Kosteleck-y 2014). While the northeast window allows for the illumination of the rock at the June solstice, Dearborn and White (1989) say that precise observation of the event is difficult at best. They suggest the use of an object suspended from pegs at the window's corners to cast a more useful shadow. When the shadow parallels the ledge on top of the rock, the solstice can be measured very accurately.

Dearborn and White (1982) describe a process by which if a plumb-bob is suspended in the northeast window its shadow will be cast upon the rock and touch its

Fig. 9.10 June solstice sunrise over the Sacred Plaza and Temple of Three windows (Author's photograph)

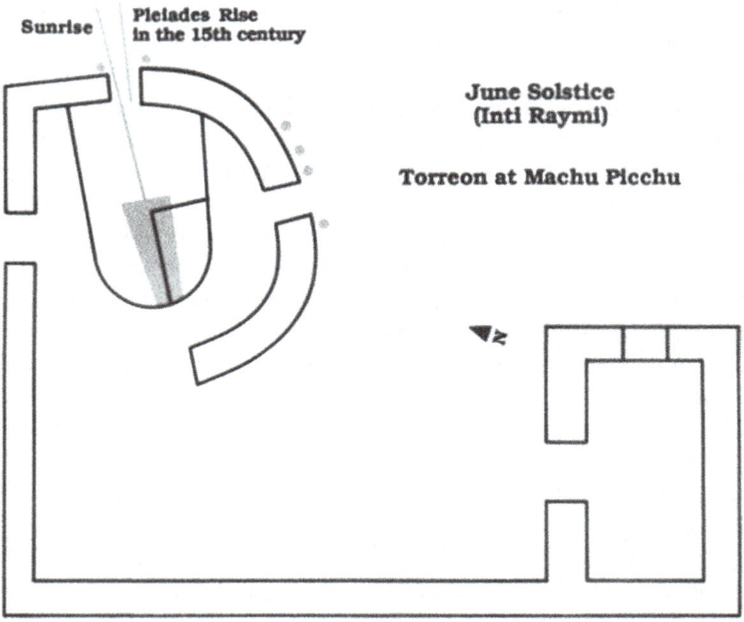

Fig. 9.11 Plan of the Torreon (from Dearborn and Schreiber 1986)

Fig. 9.12 The Torreon's northeast window within its curved stone wall crafted seamlessly atop natural rock (author's photograph)

Fig. 9.13 The Royal Mausoleum beneath the Torreon (Author's photograph)

Fig. 9.14 June solstice sunlight illuminating the boulder within the curved masonry wall of the Torreon (Author's photograph)

solstice-aligned ledge. Each day as the horizon point of the sunrise moves northward, the angle between the plumb-bob's shadow and the ledge becomes smaller until they meet on the day of the solstice. If the string is positioned carefully (which is not in the center of the window) the shadow will align with the boulder's ledge, but the ledge was not carved so that it could have been used as a direct sighting device. Further research is required to establish if this procedure was one used by the Incas.

Dearborn and White (1982, p. 253) also describe a method for using the Torreon to indicate the time of zenith passage. This procedure utilizes both the northeast and the southeast windows. Light at sunrise passes through the northeast window

Fig. 9.15 The Sun's reflection in the southern mortar at local noon at June solstice (Author's photograph)

from February through October and the southeast window from October through February. At the end of either period light from the sunrise enters both windows at the same time for a period of about five or six days. They state that the time of zenith passage could be measured by observing the illuminated areas within the Torreon at sunrise.

The Mortars

Carvings called The Mortars are located within a building in the Eastern Urban Sector (Fig. 9.15). They are two carved shallow basins situated on a floor surrounded by walls and are oriented such that on the day of an equinox mid-morning sunlight will pass thru an adjacent window (to the right of the mortars in Fig. 9.15) and cast a reflection off the water within the northern mortar basin. The two mortars are aligned directly north and south with respect to one another.

Water-filled basins might have been used for indirect viewing of the Sun and Moon. The building has been depicted without a roof (Wright and Valencia 2001). The walls surrounding the mortars are lined with niches, perhaps for religious artifacts. The Southern Cross is one of the prominent constellations in the

Fig. 9.16 Reflection of window oriented for the Sun at the time of the equinox (Author's photograph)

southern sky and therefore draws attention in the direction of south. Zuidema (1982a, p. 218) describes Inca interest in the Southern Cross and argues that in Cusco the 14th ceque of Cuntisuyu was aligned to indicate the direction of the rising of the constellation.

The window adjacent to the mortars is aligned to 075° (Fig. 9.16). The bearing from the north mortar through the window is 082° and from the south mortar it is 062°. The inclination angle from the mortars to the bottom of the window is 33° and to its top is 38°. The mortars are aligned on a 000°/180° axis and are located at S13–09.80°; W072–32.67°.

Intimachay

Also in the Eastern Urban Sector lies a cave called Intimachay (Dearborn, Schreiber, and White 1987, p. 349). Dearborn, Schreiber, and White state that this cave was constructed to observe sunrise at the time of the December solstice and the festival of Capac Raymi. A tunnel, like a horizontal light-tube, (Figs. 9.17 and 9.18) was oriented to admit sunlight to the cave for about 10 days before and after the solstice. It was aligned so as to cast a shaft of light to the back of the cave during this period. The window did not function to illuminate the cave, but instead was aligned

Fig. 9.17 The tunnel of Intimachay (Author's photograph)

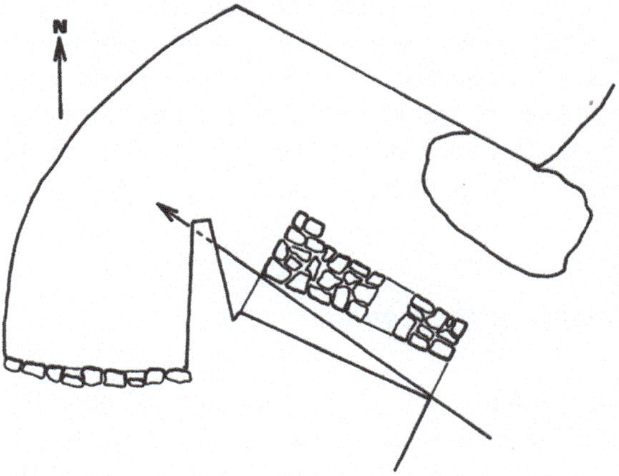

Fig. 9.18 Plan of Intimachay (from Dearborn, Schreiber, and White, 1987, p. 347)

precisely with the December solstice sunrise (Dearborn and White 1989). The view of the horizon was constrained by an interior stone that limited the field of view to 10 arcminutes (Dearborn and White 1989, p. 467). Reconstruction performed in 2006 narrowed the window and presently inhibits direct view of the horizon from the cave's interior.

Fig. 9.19 Condor stone, rock wings, and cave entrance in the Temple of the Condor (Author's photograph)

Ziolkowski, Kosciuk, and Victoria (2013) performed extensive research at Intimachay and assert that this was indeed an astronomical observatory created primarily for calendrical purposes. Capac Raymi at December solstice was a festival celebrated by the nobility that included ceremonies of passage to manhood for young Inca noblemen (Cobo, 1983 [1653], pp. 126–134). This site could have played a role in such an event.

Temple of the Condor

South of Intimachay is the Temple of the Condor (Fig. 9.19) where designers carved a head in stone while incorporating in situ rock as wings in the site's overall visual image of a condor, the creature that represented Hanan Pacha, their world above. The site's cosmological significance continued to the underworld through a system of three caves with an entrance below the boulder representing a wing of the condor. Steps downward are found at the rear of this entrance area. James Westerman (2005) and Alfredo Valencia excavated the caves in 1995, finding fissures, stairs, walls, rooms, and numerous bones of cuy (guinea pigs) which were frequently used as food in Inca ceremonies (Fig. 9.20).

Fig. 9.20 Looking out from the cave entrance over the condor stone in the direction of the anti-zenith sunrise (Author's photograph)

Zuidema (1981a) suggested that the Inca were interested in the timing of the region's anti-zenith passages of the Sun, occurring on April 26th and August 18th, the two days that the Sun is at nadir for the latitude of Cusco. Anti-zenith is discussed in Sect. 6.5. The Temple of the Condor's cave is oriented to the anti-zenith sunrise with a true azimuth of 074°, and therefore could have played a role in associated ceremonies if such festivities took place. Westerman (2005, pp. 343–344) relates that as the Sun rises on or near days of the anti-zenith its rays pass between two external structures, illuminate the condor stone, and extend beneath the boulder to the stairs at the cave's entrance (Fig. 9.21).

Huayna Picchu

Huayna Picchu was one of the major sacred features of Machu Picchu (Fig. 9.22). The primary trail to the top of Huayna Picchu leads north from Machu Picchu on its southern face, while a secondary trail begins with a double-jamb doorway at the Temple of the Moon and ascends the mountain's northern face. Near the top of the mountain is the Building of Three Windows, some limited terracing, and a tunnel, perhaps as a portal for ritual movement between the worlds of Kay Pacha and Hanan

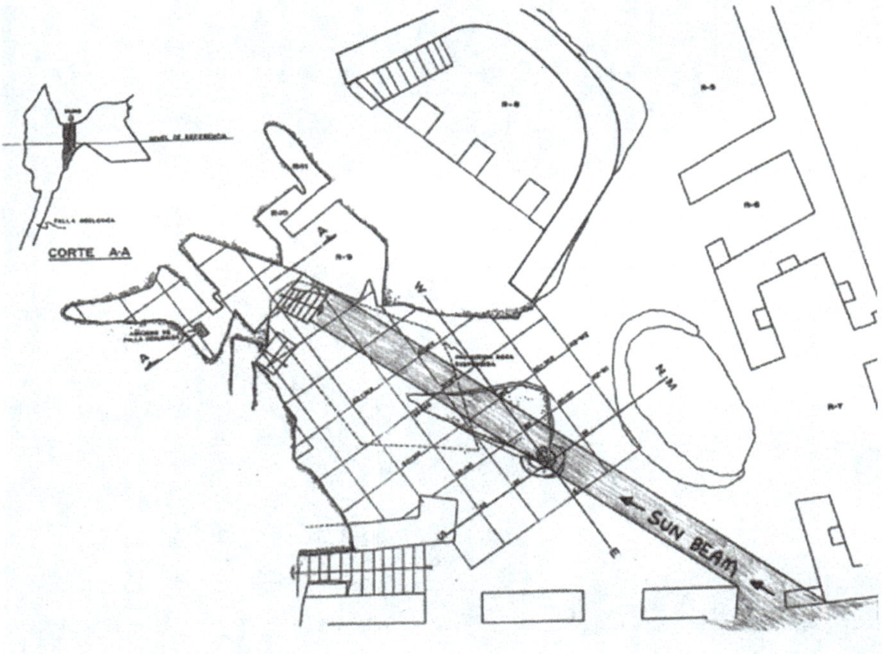

Fig. 9.21 Plan of the Temple of the Condor that depicts the path of light at anti-zenith sunrise (Alfredo Valencia, from Westerman, 2005, p. 347)

Fig. 9.22 Huayna Picchu (Author's photograph)

Fig. 9.23 Inca structures at the summit of Huayna Picchu (Author's photograph)

Pacha (Fig. 9.23). Huayna Picchu provides a commanding view of Machu Picchu and the modern switch backed road below (Fig. 9.24). The Building of Three Windows (Fig. 9.25) approximately faces December solstice sunset, the River Intihuatana, and Llactapata. The structure has been identified as a storehouse by Wright and Valencia (2001) and contains a carved stone which appears to be a replica of the Llactapata ridge. A store house may have provided supplies for elite pilgrims who had climbed the mountain. The building has been reconstructed.

Huayna Picchu's summit is located at S13–09.40; W072–32.57 and is at 2698 masl. The Building of Three Windows is S13–09.40; W072–32.76 and at 2690 masl. It faces a 244.5° bearing.

Temple of the Moon

Low on the northwest face of Huayna Picchu is a huaca referred to as the Temple of the Moon. This shrine consists of two caves, the upper (hanan) and the lower (hurin). The hanan cave is the larger and within it are five finely constructed double-jambed niches (Fig. 9.26). The double-jambs indicate this site as being significant and might have been a possible place for the storage of mummies. Three of the niches align in the approximate direction of the June solstice sunset and can also be illuminated by

Fig. 9.24 Looking down upon Machu Picchu from the summit of Huayna Picchu (Author's photograph)

the Moon. The cave is located at S13–9.09°; W072–32.78° and is at 2277 masl. Inca interest in the June solstice sunset at this site remains uncertain.

Further northeast and lower on the slope is the hurin cave which features a constructed doorway bracketed by two windows (Fig. 9.27). As with the niches in the hanan cave, the hurin cave's door and flanking windows are approximately aligned with the June solstice sunset (Fig. 9.28).

Two routes approach the site from the direction of Machu Picchu. One branches downward from the main trail between Machu Picchu and Huayna Picchu, while the other descends directly from Huayna Picchu's summit. A third lower trail passed through a gateway on its way to the river and perhaps to the River Intihuatana.

9.3 River Intihuatana

During his exploration of the Vilcabamba in 1911, Hiram Bingham located two carved rocks that he identified as intihuatanas. One of these, the Intihuatana of Machu Picchu, is one of the best-known carved rocks of the Inca world. The second intihuatana, lying deep in the Urubamba river valley to the west of Machu Picchu is visited less frequently. When examined this shrine was found to be rich with

Fig. 9.25 The view of the Urubamba River canyon and the Llactapata ridge from above the Building of Three Windows near the top of Huayna Picchu (Author's photograph)

cosmological symbolism. The River Intihuatana appears that it could be an important element of an extended ceremonial complex that combines Machu Picchu with sites on the Llactapata ridge. It is located at S13°-10.54′; W072°-33.44′ and is at 1819 masl on a hillside between PeruRail switchbacks near a hydroelectric

Fig. 9.26 Double-jambed niches in the Temple of the Moon's upper cave (Author's photograph)

Fig. 9.27 Door and windows aligned for the June solstice sunset in the lower cave of the Temple of the Moon (Author's photograph)

Fig. 9.28 June solstice sunset in the lower cave (Author's photograph)

complex (Fig. 9.2). The River Intihuatana is a multifaceted complex that was found to exhibit carved rocks, cardinal and equinoctial orientations, a zenith orientation, stairs, niches, a water source, fountains, basins, caves, a platform, a gnomon, structures and terraces.

The principle element of the shrine is a rock carved with steps and tiers (Fig. 9.29). The adjacent upslope section of the sanctuary contains two water basins aligned east-west and has an elaborately engineered water fountain that is situated over a small cave. Eastward of these granite carvings are the remains of several support structures and a tower attached to a large boulder with a second cave beneath. The area exhibits agricultural terraces, but they are presently engulfed by trees.

The significance of the River Intihuatana has become clearer since the exploration of the Llactapata Sun Temple that began in 2003. The River Intihuatana has now been identified as a major shrine (a huaca sanctuary) thought by some to be connected to Machu Picchu by two intersecting sightlines or ceques from the Llactapata ridge. The multiplicity of symbolic motifs suggests ceremonial significance at the site.

The massive carved stone lies approximately on the June solstice sunrise line as traced from the Llactapata Sun Temple across the canyon and over Machu Picchu's Sacred Plaza. The sanctuary employs east-west sightlines and adjacent to the Intihuatana is a platform that provides views of the Overlook Temple on the

Fig. 9.29 Carlos Aranibar (right) interviews Esteban Mayta, property owner, regarding his knowledge of the history of the site as they sit on the River Intihuatana (Author's photograph)

Llactapata ridge and the Intihuatana of Machu Picchu. The grounds also include several common motifs found in other Incan huacas: a fountain, basins, symbolic stairs, and caves.

The large number of huacas in Machu Picchu is reminiscent of the concentration of huacas near the Coricancha of Cusco, where 41 ceques originate. Ceques have many interpretations, one of which is that of visual alignments. A compelling orientation involves the River Intihuatana. The large granite rock contains a leveled platform that provides views of the Overlook Temple on the Llactapata ridge and the Intihuatana stone of Machu Picchu. The location of the rock is critical for its viewscape; if it had been located a few meters to the south, the view of the Overlook Temple would have been blocked by canyon walls. The Machu Picchu Intihuatana is clearly visible (Figs. 9.6, 9.7, and 9.35) and the base of the River Intihuatana's carved stone is oriented approximately in its direction.

The River Intihuatana lies just above the Urubamba River hydroelectric compound that is situated 5 kilometers southwest of Aguas Calientes. The primary carved stone is located approximately 20 meters downslope (north) from the upper PeruRail tracks entering the area. About 30 meters below the sanctuary platform is the mid set of tracks of the switchback and then the slope continues downward to a final set of tracks and across to the hydroelectric compound and the river.

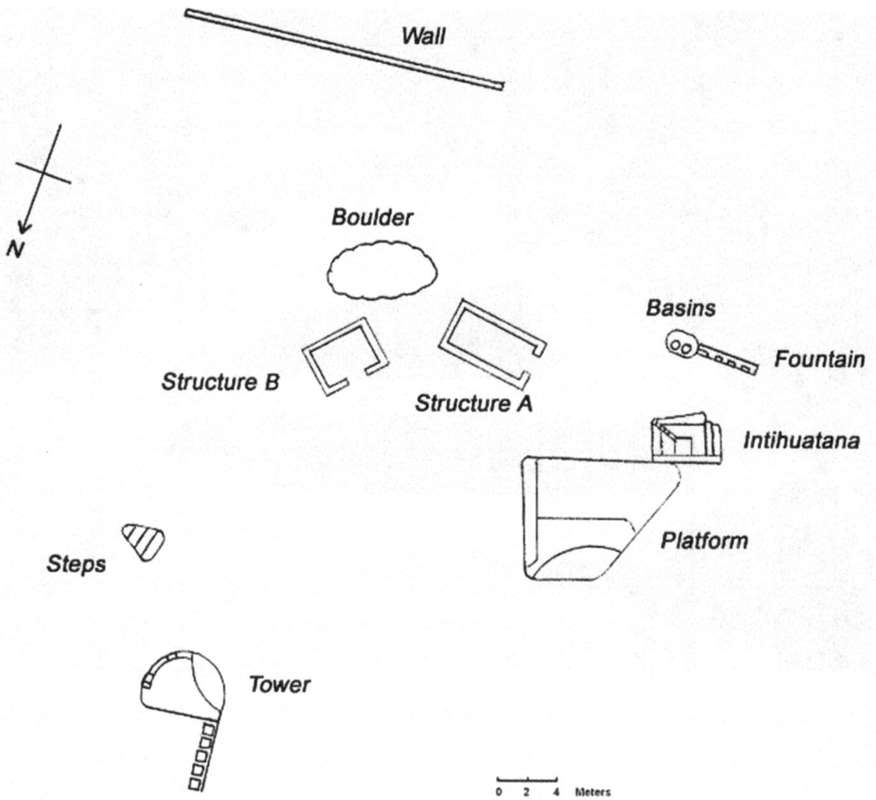

Fig. 9.30 Ground plan of the River Intihuatana sanctuary (author's diagram)

Fig. 9.30 is a ground plan of the sanctuary's major features. The upper set of tracks run left to right above this diagram and from it a trail descends the slope, passing just to the left of the basins before reaching the Intihuatana and platform. The walls of the support structures begin five meters to the east of the Intihuatana with Structure A. possibly having served as quarters for the huaca's attendants and Structure B perhaps functioning to house stores. Structure B is four meters from Structure A. Immediately upslope from the two structures is a very large unimproved boulder. Approximately two meters upslope from Structure A begins a terrace that extends until reaching a wall that runs nearly eastward for 22 ½ meters along the hillside.

Fifteen meters northeast of Structure B and set into the ground is a set of three steps (Fig. 9.40). Further downslope exists the site's remaining major structure, a circular tower built into the side of the hill (Figs. 9.41, 9.42, and 9.43). Five functional stairs descend from the base of this tower, ending near the switchback's mid tracks.

Fig. 9.31 The River Intihuatana as viewed looking downslope to the north (Author's photograph)

Extending to the east and encompassing the majority of the slope between the upper and mid sets of tracks is a group of multiple andenes. These agricultural terraces may have been used in partial support of the huaca's attendants, but currently are hidden by a thick expanse of trees. This likely was clear when the complex was occupied some 500 years ago.

Intihuatana

The site's primary feature is the Intihuatana (Figs. 9.31, 9.32, and 9.33), a somewhat worn, but finely carved stone situated at the sanctuary's western boundary. Its dimensions overall are 4.27 meters along the flat northern face by 3.20 meters wide. The tiers get increasingly smaller as they rise. The middle tier measures 2.17 meters by 2.14 meters and the top tier 1.50 meters by 1.70 meters. On the east side of the top tier is an intermediate level measuring 48 cm by 1.70 meters and both are adjoined by a set of descending steps too small to serve any necessary function as they are situated. There appears at present to be three steps, but the stone has been subject to enough erosion to make the original number uncertain. The northern face of the Intihuatana is a flat wall angled from the west toward the top. Its eastern side

Fig. 9.32 The River Intihuatana from the east (Author's photograph)

Fig. 9.33 The River Intihuatana's western side (Author's photograph)

Fig. 9.34 A magnified view of the locations of Machu Picchu's Intihuatana and Sacred Plaza as seen from the River Intihuatana (Author's photograph)

is also angled, but the surface is broken, thus masking its original shape. The top of the northern face is flat and includes remains of what was perhaps a gnomon. The base of the shrine is partially covered by soil and the western edge of the area encompassing the Intihuatana and the fountain drops vertically along a sheer stone face. The Intihuatana may also have had a higher portion that could have been broken by such as a landslide.

The lines of the River Intihuatana have been rotated from the site's predominant east-west orientations to point in the direction of Machu Picchu's Intihuatana, about 044°. The line of sight is clear between the two intihuatanas and they may easily have been viewed from one another (Fig. 9.34).

Platform

Adjacent to the northeast corner of the Intihuatana lays an expansive stone platform (Fig. 9.35). This construct is larger than the Intihuatana with overall dimensions running 7.64 meters wide and 10.38 meters long. An angled edge extends from the Intihuatana on a bearing of 20° for 8.80 meters. The platform incorporates some carved tiering and ends downslope in a descending vertical masonry wall.

Fig. 9.35 Tiers of the platform adjacent to the River Intihuatana (Author's photograph)

Fountain

One of the more intriguing areas found on the site is a complex incorporating several common huaca motifs: a fountain, two basins, and a cave. The fountain structure (Fig. 9.36) is situated 3.84 meters upslope from the Intihuatana and spans 5.60 meters at its extremes. The face of the fountain is oriented approximately on a 090/270° axis and was designed to receive water from the east into the channel shown in Fig. 9.37. A ledge was carved 1.56 meters below the top of the fountain and worn examples of seats or shelves remain sculpted to the west. The channel was engineered to distribute water to each of the four outlets on the fountain's face. The outlets measure 16.5 cm by 9 cm and are spaced 61 cm apart. Within the channel a small baffle was constructed at each outlet to ensure an even dispersion of water flow through that opening. The fountain is now dry but would have once been fed by an upslope spring.

Basins

Immediately adjacent to the eastern end of the fountain is a boulder with two large carved basins (Fig. 9.38). The boulder is 3.20 meters from the Intihuatana and measures 2.80 meters wide by 1.55 meters deep. The basins are aligned east-west

Fig. 9.36 Carlos Aranibar measures the fountain water outlets (Author's photograph)

with the east basin measuring 66 cm wide by 33 cm deep. The west basin has dimensions of 56 cm by 33 cm and the two are 13 cm apart.

The basins are aligned on an axis between the Overlook Temple of Llactapata and Machu Picchu Peak. This orientation allows possible use for observing reflections of the sunset at the Overlook Temple. The east-west orientation of the basins is consistent with the overall organization of the site.

Caves

A small cave exists within the fountain-basin complex (Fig. 9.39). This orifice extends beneath the fountain with its opening situated between the fountain and the boulder with the basins. The cave has ample room for an attendant or priest to function or could possibly have been used to store mummies. Above the cave entrance and carved into the boulder is a set of three symbolic stairs, perhaps here as well representing transition between the underworld and the world of the here and now. The Incas felt caves to express deep connections with the forces of nature.

A second cave was described by Esteban Mayta, a local informant and the landowner, and exists beneath the base of the boulder that forms part of the tower constructed on the sanctuary hillside (Fig. 9.43). Mayta reported that the caves were deep enough to occasionally entrap animals.

Fig. 9.37 Carlos Aranibar and the canal with carved baffles designed to produce equal water flow in the four fountain outlets (Author's photograph)

Structure A

Structure A has rectangular stone walls and lies to the east of the Intihuatana, 4.72 meters from Platform P. This appears to have been a building able to serve as housing for the huaca's priests or other caretakers. The length of the structure is 6.40

Fig. 9.38 Two water-filled basins aligned on an east-west axis (Author's photograph)

Fig. 9.39 Author entering the cave below the fountains (Author's photograph)

meters and its width is 4.45 meters. Its western wall contains a door 1.04 meters wide. The long walls are oriented approximately 100/280°, with the side walls about 010/190°.

Structure B

Structure B is situated 3.58 meters east of Structure A and is more nearly square in shape with sides 3.83 meters long and 3.81 meters wide. A 97-cm door is situated at the center of the north wall and the structure is oriented approximately 050/230° and 130/310°. This building might have been used as a storehouse.

Boulder

Immediately upslope of Structure A and Structure B is a very large, unimproved boulder. It is curious that such a significant rock was left uncarved.

Steps

At a distance of 14.9 meters northeast of Structure B lies a small set of isolated stairs embedded in the ground (Fig. 9.40). Their positioning is of little practical use, but they exhibit the three traditional steps of Inca cosmology.

Tower

Further downslope built into the side of the hill is a round tower (Figs. 9.41; 9.42, and 9.43). The stones of an eastern circular wall adjoin a large in situ boulder that was used for the tower's southwestern side. The northern face is presently open. The eastern wall contains a door oriented to 90° and the door is 2.10 meters opposite the in situ boulder. The doorway is 1.28 meters high, 60 cm wide at the lintel, and 74 cm wide at its base. Approximately equidistant between the door and the rock is an internal niche with an upper lintel at the same height as that of the door. The niche is 56 cm high, 29 cm wide at the lintel, and 34 cm wide at its base. A trapezoidal upper window is situated 78 cm above the niche and slightly offset toward the door. The window is 55 cm high, its upper surface measures 28 cm and its base measures 40 cm. The window is oriented to 125°.

A stone wall holds back the earth outside of the doorway's entrance on the east side of the tower and continues downslope for 8 meters. Another stone wall retains

Fig. 9.40 Three symbolic steps on the huaca sanctuary grounds (Author's photograph)

Fig. 9.41 Looking down into the tower from above. The doorway is to the left of center; the entrance to the cave is immediately below and adjacent to the free stone (Author's photograph)

Fig. 9.42 The towers' eastern side (Author's photograph)

Fig. 9.43 A curved masonry wall with a door, niche and window adjoins an in situ boulder in the River Intihuatana tower; the cave entrance is to the right of the stone (Author's photograph)

the soil on the downslope of the north side. On the tower's west side a stone staircase with five steps descends the slope on a 000° bearing toward the railroad. The top of the tower is open.

Zuidema (1981b, p. 340–341) describes the Sunturhausi in the Hanan Haucaypata plaza in Cusco as being a circular building with windows, a high roof, and a mast on top. He identifies it as a site for visual observations of the Sun and perhaps a marker of the zenith Sun with similar use and stature as the Coricancha or the Muyucmarca tower of Sacsahuaman. He mentions them all as belonging to the same category of "temple of the Sun." The utility of the River Intihuatana's circular tower would be more effective for zenith solar observations than for sunrise and sunset positions limited by the high horizons. A circular tower reaching to the sky while sitting atop a cave may have served to symbolize and facilitate strong connections between this world and those of the heavens and the underworld.

Andenes

A system of andenes extends along the hillside to the east of the sanctuary and are oriented generally east-west. The terraces may have been used both to control erosion and also for agriculture in support of the sanctuary's attendants. This area likely would have been clear in the time of the Incas, but presently is embedded in vegetation.

Design Similarities

The relationship between the River Intihuatana and Machu Picchu is evident in certain similarities found in architectural style and stone carving common to both sites. The tower found near the River Intihuatana exhibits similarities with the Torreon of Machu Picchu. Both share plans that incorporate walls of curved masonry that include straight and open sections. The masonry of each adjoins in situ rock used to form the remainder of the structure and both exhibit caves at their bases. They each employ windows and niches and both exhibit solar orientations, the Torreon for June solstice sunrise and the River Intihuatana tower for sunrise at equinox (Figs. 9.14 and 9.42).

The carved stone of the River Intihuatana exhibits similarities with a carved stone near the "Guardhouse" in Machu Picchu's Upper Agricultural Sector. The stone is located on a terrace and is known as the "Ceremonial Rock" (Fig. 9.44). Both it and the River Intihuatana (Fig. 9.31) share several distinct features. They are each carved from granite with multiple tiers or shelves and have upper ceremonial platforms, as well as steps sculpted into one corner. Both carvings seem to follow the same philosophy of design with similar craftsmanship and give further indication of the River Intihuatana's close association with Machu Picchu.

Fig. 9.44 Carved steps and tiers in the Ceremonial Rock of Machu Picchu's Upper Agricultural Sector resemble those of the River Intihuatana; scattered stones carried into Machu Picchu are visible in the foreground (Author's photograph)

Numbers of small-elongated stones, many about 30 cm by 15 cm in size, are arranged in upright groups around the Ceremonial Rock. The stones are andesite, limestone, and metamorphic rocks carried in from other regions. Some are rounded river-shaped rocks. Ruth Wright and Alfredo Valencia suggest that "river rocks symbolically bring the sacred river to the mountain site" (Wright and Valencia 2001, p. 8). This may have been the case but the diversity of rocks more likely indicates that they were ritual offerings/burdens carried and placed by visitors at a shrine requiring this activity upon arrival at Machu Picchu. Modern Quechua travelers carry small stones to the top of mountain passes to leave as offerings to the Apus, in this case perhaps was done for Huayna Picchu. The Ceremonial Rock is the only huaca placed outside the walls of the main section of Machu Picchu. In a manner similar to the social differentiation that occurred at the Sanctuary of Isla del Sol (Bauer and Stanish 2001), these offerings may have been carried in and placed there by pilgrims who were not allowed to pass through the main gate of Machu Picchu. Potsherds of chicha vessels have also been found in this area (Wright and Valencia 2001).

The carved Ceremonial Rock (Fig. 9.44), River Intihuatana (Fig. 9.31), Machu Picchu's Intihuatana (Fig. 9.7) and the Rumihuasi at Saihuite (Figs. 7.78 and 7.79) have similar sizes and are carved on all sides. The Intihuatana at Pisac is of an appropriate size but is carved on top and enclosed by masonry walls (Fig. 8.23). The Intihuatana of Tipon (Fig. 7.64) is an exception in that it consists of several unimproved boulders atop a rock platform overlooking the site's terraces and horizon. The similarity of carving of Machu Picchu's Intihuatana, the River Intihuatana, the Ceremonial Rock, and the Rumihuasi, such as multiple sets of steps with different scales, suggests that these huacas shared common symbolic meaning and function.

Access

Access to the River Intihuatana may have been by a trail leading down from the double-jambed doorway of Machu Picchu's Temple of the Moon. That route would have required a bridge to cross the Urubamba River from the Inca trail on the north side of the river. A trail leading downward from the southwestern terraces of Machu Picchu leads directly to the Intihuatana, without the need of crossing the river. The trail from the River Intihuatana to Llactapata may be similar to the modern one, proceeding downstream to the confluence with the Aobamba and then up that river for a short distance before turning and ascending the Llactapata Ridge.

9.4 Llactapata

The 2003 exploration of Llactapata, which overlooks Machu Picchu from a distance of 5 km (Fig. 9.2), was reported at the Oxford VII conference (Malville, Thomson, and Ziegler 2006). Llactapata is located at S13°-11.12′; W072°-35.08′ and is at 2740 masl. It is a multi-faceted complex that was found to exhibit light and shadow effects, solsticial orientations, Pleiades orientations, niches, double-jambed doorways, channels, and structures.

A remarkable feature of Llactapata is the Sun Temple of Sector I (Figs. 9.45 and 9.46), which has an orientation similar to that of the Coricancha of Cusco. The courtyard in front of the double-jambed doorway of Sector I contains a stone-lined channel (Figs. 9.47 and 9.48) leading from the doorway toward the Sacred Plaza of Machu Picchu, in the approximate direction of the June solstice sunrise and also the approximate direction of the heliacal rise of the Pleiades. The River Intihuatana lies in the canyon below. The position of the sunrise could have been predicted by a recess on the jagged horizon.

Llactapata was most likely an integral part of the cosmological orientations of Machu Picchu. A photo taken from Machu Picchu's Intihuatana (Fig. 9.7) points to the location of Llactapata's Sun Temple on a ridge, which is surrounded by over 100 other structures, still engulfed by the cloud forest.

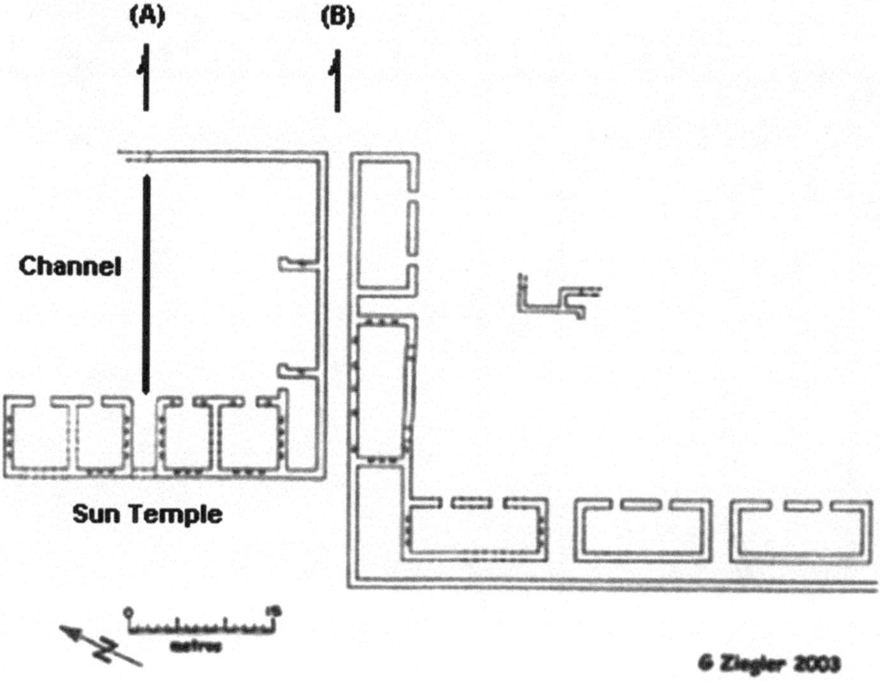

Fig. 9.45 Plan of Llactapata Sector I. The channel from the Sun Temple's primary door is aligned for (**a**) the Machu Picchu Sacred Plaza, the June solstice sunrise, and the rise of the Pleiades; the adjacent corridor is aligned for (**b**) the June solstice sunrise and the rise of the Pleiades (Gary Ziegler, modified from Malville, Thompson, and Ziegler 2006, p. 330)

June Solstice Sunrise

The rugged Andean peaks surrounding Machu Picchu and Llactapata provide ample reference for marking solar horizon events and natural features may have been used to identify sunrises or sunsets from either location. The Llactapata solstice sunrise over Machu Picchu and its close alignment with the water channel extending from the Sun Temple was photo-documented (Fig. 9.49). A similar sunrise orientation was observed from an adjacent 33-meter corridor, which is also aligned in the approximate direction of the Pleiades following its heliacal rising. It has been proposed that rooms of the Coricancha of Cusco were constructed so as to frame the rise of the Pleiades (Zuidema 1982a, pp. 212–214) (Fig. 6.7). In like fashion the corridor adjacent to the Llactapata Sun Temple appears to have been constructed to frame this event.

Water or other liquid offerings would have had to have been carried up to the double-jambed doorway at the Sun Temple to be poured into the channel as an offering to the solstice Sun. Reinhard (2002) suggests the importance of an Incan water cult at Machu Picchu. The mythical and ritual connections between the Sun,

Fig. 9.46 Sun Temple of Llactapata (authors's photograph)

Fig. 9.47 Stone-lined channel from Sun Temple's central doorway (Author's photograph)

Fig. 9.48 The view toward Machu Picchu from the primary door of Llactapata's Sun Temple with the stone-lined channel in the foreground (Author's photograph)

Fig. 9.49 June solstice sunrise from Llactapata's Sun Temple. The water channel is below the camera tripod

sacred mountains, the deity Wiracoca, the Vilcanota/Urubamba River, and the Milky Way (Urton 1981a) may have been especially powerful there. Cosmological tradition has it that the Sun is revitalized nightly from the Vilcanota River, which becomes the Urubamba River before reaching Machu Picchu (Urton 1981a, pp. 68–69). According to these traditions, the Sun moves underground along the Vilcanota River during the night, as it is refreshed from those waters, to be born again from the river. At the time of the June solstice the river is low and does not have enough water to refresh the Sun, hence its relative dimness. The pouring of water into the Llactapata channel may have symbolically represented the feeding of the Sun during the dry season.

Pleiades Rise

A 33-meter-long open corridor lies east of and adjacent to Llactapata's Sun Temple (Fig. 9.50). Similar to the Sun Temple, the corridor is orientated toward both the June solstice sunrise as well as the nearby point of the heliacal rise of the Pleiades (Fig. 6.17). The corridor, which is 33 meters long and 2.5 meters wide, is oriented on an azimuth of 063.5°. When viewed from the corridor's rear, a field of view is created extending 4.3° along the horizon. In A.D. 1500, the June solstice Sun rose

Fig. 9.50 June solstice sunrise from the 33-meter-long corridor (Author's photograph)

at an azimuth of 064.3° while the heliacal rise of the Pleiades ranged from 064.9° to 065.6° (Malville, Thomson, and Ziegler 2006).

The Pleiades were of great importance in Inca astronomy as the Incas found them useful in predicting and planning for harvests. The stars in this prominent grouping were viewed with regard to their relative brilliance. Orlove, Chiang, and Cane (2000) argue that a bright appearance by the Pleiades indicated a future of ample rain with a correspondingly good harvest. Correspondingly, a dull appearance (caused by atmospheric obscuration) indicated that there would be drought in the months to come. These simple observations became a means of anticipating the effects experienced during an El Niño year in a manner still used by Andean farmers today (Orlove, Chiang, and Cane 2000, p. 68; Sect. 6.11).

Linked Ceremonialism Between Llactapata and Machu Picchu

Sun worship was the official religion of the Inca Empire, with the royal family claiming descent from the Sun. Their legendary ancestor, Manco Capac, was said to have used a suit of gold and a golden disk to reflect light of the Sun (Hemming 1970).

The account by Cristobal de Molina of ceremonies at Inti Raymi may also provide some insight for possible ceremonies that could have occurred at the Llactapata Sun Temple at the time of June solstice (Hemming 1970). The 33-meter corridor would have allowed for a procession of pairs of people facing the rising Sun. At the head of the column, the Inca wearing his golden attire could have been a brilliant object at sunrise.

The Coricancha means the "golden enclosure." The sides of the temple facing sunrise were covered with gold by at least 700 plates averaging 2 kgs each, amassing more than 1400 kgs. Additionally, there was a band of gold with a width of 20 cm that ran along the entire building at roof level (Hemming and Ranney 1982).

If the Llactapata Sun Temple was similarly plated, it would have been a brilliant sight when reflecting the light of the dawn Sun at the time of the June solstice. Since Llactapata is at a higher elevation than is Machu Picchu, it receives illumination first on the morning of the solstice.

To demonstrate the possibility for such a ceremony I left a field assistant positioned at the Sun Temple the day prior with a 15 cm by 20 cm mirror. Sunrise occurred at Llactapata at 06:41 the next morning, and on cue the assistant captured the Sun in the mirror and reflected its light toward Machu Picchu's Sacred Plaza. The reflections from 5 km across the canyon were prominent and continued through the Sacred Plaza's sunrise at 07:17 and beyond. The 36-minute period between the Llactapata and Machu Picchu sunrises would have afforded ample time for special ceremony celebrating the morning solstice Sun. A magnified view of the ridge (Fig. 9.51) clearly shows the Sun Temple, visible due to clearing performed by the Peruvian government. This demonstration showed only the reflection from a handheld mirror. Larger medallions worn by the emperor or plates

Fig. 9.51 Reflection of the June solstice sunrise at the center of the Llactapata Sun Temple as seen from Machu Picchu's Sacred Plaza (Author's photograph)

mounted on the temple could have produced a more brilliant effect. It must be pointed out, though, that there is no evidence in the chronicles that this type of ceremony took place.

9.5 Summary

The Incas encoded many examples of astronomical and cosmological significance at Machu Picchu and the surrounding region. The Sacred Plaza and Torreon are aligned with a solar axis extending from the horizon point of the June solstice sunrise to that of the December solstice sunset. The Intihuatana is one of Machu Picchu's most prominent features and can be said to look out over the December solstice sunset. The mortars may have been used for indirect observations of the Sun and Moon or to reflect sunlight entering through a window on days of equinox. The Intimachay was designed to observe sunrise at December solstice and the Temple of the Condor has a potential sunrise alignment for days of the anti-zenith Sun. The Building of Three Windows near the top of Huayna Picchu is oriented to the December solstice sunset and features of the Temple of the Moon lower on Huayna Picchu's side align for sunset at June solstice.

The River Intihuatana lies at the junction of two potential solar axes, a northeast-southwest solsticial one and another with an east-west equinoctial alignment. The features of the River Intihuantana sanctuary are also oriented predominantly east-west.

The River Intihuatana is comprehensive in its utilization of the many motifs found in other Inca huacas. Features such as stairs, fountains, basins, and caves are common, but it is less frequent when they are all found incorporated together. Ritual stairs express movement from the underworld to the Earth to the heavens and are frequently associated with these three realms. Three examples are found within the River Intihuatana sanctuary: standing alone on the grounds of the complex, situated above the cave entrance between the basins and the fountain, and carved into the Intihuatana itself. None of these were carved in a way that suggests any practical need and the steps above the cave are totally non-functional.

The Incas designed fountains to be both practical and ceremonial and they also likely served to facilitate the life-force energizing effects of camay. Carved basins were often used to hold gifts and liquid sacrifices and also might have been employed to observe reflections of the Sun or Moon.

The River Intihuatana's circular tower reaching to the sky while sitting atop another cave may have served to symbolize and facilitate strong connections between the world of the present and both the gods of the heavens and the forces of the underworld. In comparison, it seems unusual to have a tower in a location such as at the bottom of a canyon where the visual horizon is limited.

Llactapata and its extensive array of structures suggest an even greater significance for the Machu Picchu region and also that Pachacuti may have created a ceque system here not unlike the one he instituted surrounding Cusco. A stone-lined channel in front of Llactapata's Temple of the Sun is aligned toward Machu Picchu, the June solstice sunrise, and the approximate heliacal rise of the Pleiades. An adjacent corridor points specifically to the rise of the Pleiades and that of the June solstice Sun. The alignment between the Temple of the Sun, the River Intihuatana, and Machu Picchu's Sacred Plaza lies on a solar axis from June solstice sunrise to December solstice sunset. Llactapata's Overlook temple is aligned across the River Intihuatana to Cerro Machu Picchu on an equinoctial axis. It has been hypothesized that the Emperor may have conducted June solstice sunrise ceremonies at Llactapata where he reflected the light of the rising Sun to onlookers at Machu Picchu's Sacred Plaza, but there is no evidence that supports this possibility.

Chapter 10
Discussion, Findings, and Conclusion

Fig. 10.1 Women with Dogs (watercolor by Jessica Gullberg)

10.1 Discussion

The ancient Incas practiced horizon astronomy in the Peruvian Andes through positional observations of the rising and setting Sun on days of ceremonial and agricultural significance. They also observed risings for celestial objects such as the

© Springer Nature Switzerland AG 2020

S. R. Gullberg, *Astronomy of the Inca Empire*, Historical & Cultural Astronomy,
https://doi.org/10.1007/978-3-030-48366-1_10

Pleiades. A majority of the sites examined involve attention to the solstices. Solar orientations are preeminent and have been the primary focus in this book.

The pillars above Q'espiwanka lie on a sightline to the June solstice sunrise when viewed from a large boulder at the palace below. The Sacred Plaza of Machu Picchu, the River Intihuatana, and the Llactapata Sun Temple lie approximately along an axis established by the June solstice sunrise and December solstice sunset. The River Intihuatana is a classic huaca with a site containing caves, fountains, steps, and platforms. The Temple of the Moon at Machu Picchu exhibits approximate orientations for the June solstice sunset. The huaca at Choquequilla contains some very fine carvings and the opening of its cave aligns for the December solstice sunrise. The site also has been mentioned with regard to the Moon. Chinchero includes both cardinal and solsticial orientations. The carved rock huaca at Kenko Grande is rich with astronomical orientation and its neighbor at Lacco features even more. Huaca 44, with gnomons indicating orientations for solar horizon events, may have been used to determine the positions of solsticial and equinoctial sunrises and sunsets. Insight provided in the first six chapters enhances appreciation for what is presented in Chaps. 7, 8, and 9.

Cusco Region, Tipon, and Saihuite

Chapter 7 includes examples such as the "Eyes of the Puma" at Kenko Grande. Here the Incas demonstrate a considerable knowledge of horizon astronomy and the degree of the creativity that they were capable of in the development of their shrines. On the top of the outcrop they carved two gnomons that form the puma's eyes and then created a fissure in the nearby wall that allows light to fall upon those gnomons at the appropriate time during the June solstice sunrise in a manner that completes the "puma" visual effect, as shown in Fig. 7.6.

With its many carvings and motifs, Kenko Grande was truly a most important huaca. Near the puma is the kenko, or zigzagged channel, from which the huaca appears to derive its name. Such channels appear at several locations in the empire for suggested purposes of offerings and divination. A likely offering was chicha, or maize corn beer, which is still homebrewed in traditional fashion and can be available at certain bars and restaurants, especially in rural villages. The drink was common in Inca times and likely held significant spiritual value. "These women made large amounts of high-quality *chicha* which was for offerings to the gods and for the priests to drink" (Cobo 1990 [1653], p. 174). Other fluids could be used in these channels as well. Certainly, water with its sacred connotations was incorporated into ritual, and blood, such as of the llama, could have been used at times of sacrifice. Many channels incorporated multiple paths and the specific flow of the liquid was likely used for divination.

The chamber within Kenko Grande exhibits traditional motifs of caves, niches, and steps. The primary altar is finely carved and may have served in one or more types of ceremonial functions. The tube in the cave's upper northwest corner admits light that could be reflected from plates in a niche to illuminate the entire chamber. It facilitates a special effect of light and shadow on the ceremonial steps adjacent to the primary altar, as depicted in Figs. 7.11, 7.12, and 7.13. The huaca has an east or west opening at respective ends of its cave, with the southeastern one oriented to receive the light of the equinox sunrise, and the one to the northwest admitting illumination from the equinox sunset. The secondary altar, situated near the southeast entry, is illuminated at dawn during the time of the March and September equinox sunrises.

Lacco exhibits prominent examples of the Incas' passion for using light from the Sun or Moon in their shrines and ceremonies. The huaca was finely carved across its top and still shows many examples of Inca sculptures such as seats, steps, and animals. It is the caves, however, that truly demonstrate the prowess the Incas had for solar orientation. Each of the three has an altar that the Incas deemed important enough to illuminate at certain times by the Sun or Moon. The exact ceremonies performed remain subject to interpretation, but the ability to utilize solar or lunar light while conducting them is clear. The Northeast cave's altar is fully illuminated in the early morning on days approaching, during, and following the June solstice. The alignment of the cave opening's center with the sunrise point on the horizon is striking, as was shown in Fig. 7.25. The Sun's rays are cast directly upon the altar and reflections brightly illuminate the rest of the chamber. Two large steps are situated immediately in front of the cave and may have been a part of associated ceremonies. In front of the steps are two large separate thrones that also are oriented toward the solstice sunrise. The thrones would have been excellent locations for observing the rising Sun.

Lacco's Southeast and Southwest caves incorporate specifically oriented light-tubes to admit the rays of the Sun or Moon. The Southwest cave is the smaller of the two and also has a smaller altar. Its light-tube's alignment with the general path of the Sun was shown with the crescent Moon in Fig. 7.30 and light will be admitted when the Sun or Moon crosses the sky at an angle of 70–75° above the horizon. The Southeast cave, also known as the Temple of the Moon, appears to have been the most prominent of the three as evidenced by the degree of workmanship in its sculpture, complete with fine carvings of a puma and snake near its entrance and the highly polished altar within its inner chamber. The altar is of an appropriate size and height for ceremony and sacrifice and is brilliantly illuminated by the Sun when it approaches 90° above the horizon near the time of zenith, as was shown in Fig. 7.34. The zenith orientation of this light-tube is important due to the significance that zenith passage plays in the Inca calendar. There is a deep chamber adjacent to the altar that gives the appearance of entry into the underworld, a cosmological connection that is common with many Inca caves.

Huaca 44, the huaca located between Kenko Grande and Lacco with two gnomons that exhibit orientations indicating horizon positions of the solstice and equinox solar events, is especially interesting. It is intriguing to find a construct that includes orientations for all primary solar horizon positions, and these are shown in Figs. 7.38 and 7.39. This site may as well have had other functions but, for reasons yet to be determined, watching solstice and equinox sunrises and sunsets appears to have been important here. The many seats carved upon the huaca underscore this relationship as there is at least one oriented for each of the associated Sun risings or settings, with the exception of the June solstice sunset where that part of the outcrop has eroded extensively. The full extent of the Incas' purpose for this huaca remains to be seen.

The separation of elites and non-elites for the viewing of special solar events is an interesting concept that is among the possibilities at Tipon and Q'espiwanka. Bauer and Stanish (2001) suggested that the Incas separated such viewings at the Island of the Sun in Lake Titicaca. Both Tipon and Q'espiwanka also appear to be oriented for class division regarding solar ceremonies. At Tipon there is a double-jambed door along the trail leading up the hill to the Intihuatana complex. It appears that Inca rulers desiring to indoctrinate the masses may have created separate viewing areas for non-elites aside from those of the elite. In each case alignments were such that the non-elite viewers could see the Sun rise or set over the elites, perhaps over the ruling Inca himself. As the son of the Sun, this effect would certainly be a valuable part of the state-sponsored indoctrination received by pilgrims. The same could have occurred at Q'espiwanka where non-elites were barred from the main palace grounds by a triple-jambed entranceway but could view the pillars on the ridge and the June solstice sunrise over the palace from outside its south wall.

The Sacred Valley

As shown in Chap. 8, the many structures at Chinchero were constructed with cardinal direction alignments and a question that remains is by what means did the Incas determine them? It is possible to establish such directions by shadow at the times of equinoxes. Otherwise a gnomon shadow plot could be used to establish true south. No matter what method was used, the end result is impressive, as depicted in Fig. 8.6.

Chinkana, Mesakaka, and Kondorkaka are all aligned north and south with regard to one another. While it is unlikely that Chinkana was moved, the other two huacas are small enough to have had their positions adjusted. Chinkana and Titikaka are shown in Fig. 8.19 to be aligned approximately on the axis of the June solstice sunrise and December solstice sunset. Both rocks would seem to be immovable, so how were the Incas fortunate enough to have these two major outcrops align with this most important axis? Regardless, they used the alignment to their advantage with the orientation of the triangular basin and the balcony on Chinkana, as well as the hole on the upper surface of Titikaka. Paternosto (1989) talks much about Inca textiles and it would seem possible that they may have hung a banner from a pole set into this post-hole. The banner, or flag, could then have been a visual part of the alignment of the rocks for the June solstice sunrise or the December solstice sunset. As viewed from Chinkana's balcony, a large banner atop Titikaka could help accentuate the ritual setting of the December solstice Sun.

Pisac's most prominent feature is its intihuatana, as displayed in Fig. 8.23. The name and meaning of *intihuatana* is a matter of debate and is now thought to be somewhat of a misnomer applied in more recent times. Traditionally these may have been "hitching posts of the Sun," or the term simply may have been given to them by local villagers in later years. The stones are generally among the most impressive of the sites they occupy, and the one at Pisac is no exception. It is a carved huaca with a gnomon on top. Unlike other intihuatanas, though, this one is enclosed by masonry walls in the shape of the letter "D." At Tipon the intihuatana is different as it has not been carved and instead consists of several unimproved boulders that are gathered together on a prominent point of the site. The intihuatanas at Machu Picchu and the Urubamba River are both finely carved on all four sides. The Machu Picchu Intihuatana is located on high ground, while the River Intihuatana is low in a river valley. The River Intihuatana also shares features in common with stones such as the Ceremonial Rock of Machu Picchu and the Rumihuasi Stone of Saihuite. Neither of these rocks are known as intihuatanas, but they are quite similar in design to the one by the Urubamba River. Perhaps there is no definitive answer as to whether some of these sites actually constitute intihuatanas in the traditional sense, if they should be considered as such, or even what intihuatana really means.

The pillars above Q'espiwanka in Fig. 8.27 are noteworthy examples of a piece of Inca history. Spanish chroniclers recorded solar pillars on the horizons of Cusco, but all were eradicated, presumably in the Catholic purge of Inca idolatries. Their locations were recorded, but no remains have been found. Accounts speak of them being used to monitor solar positions, but there is no extant evidence to this effect. The pillars near Urubamba give credibility to the reports in Cusco and enable the study of this form of horizon astronomy first-hand. Two towers survive on the Cerro Sayhua ridge and demonstrate the feasibility of this method to mark the June solstice sunrise. The alignment supports the Bauer and Dearborn (1995) hypothesis with regard to these structures as well as Niles' (1999) suggestions regarding the prominence of the large granite boulder and the adjacent platform that were the central features of Huayna Capac's palace grounds. The utility of the boulder in Fig. 8.25 and a no longer extant platform as the focal point of the plaza is supported by this orientation and, to a lesser extent, by potential alignments with natural horizon features for the December solstice sunrise and the June solstice sunset. The remains of a wall on top of Cerro Pumahuachana and the three pillars at the peak of Cerro Unoraqui remain enigmatic and inspire many questions as to their meaning and purpose. Ollantaytambo exhibits multiple astronomical orientations for the solstices and the zenith Sun.

Machu Picchu Region

Chapter 9 demonstrated how rich Machu Picchu is with astronomical symbolism. The orientations exhibited for the major solar axis of the June solstice sunrise/December solstice sunset are especially noteworthy. Within Machu Picchu, the Torreon and the Sacred Plaza each have alignments associated with these horizon events and the Sacred Plaza, as shown in Fig. 9.8, might have formed one end of what could have been an intentional ceque that included the River Intihuatana and the Llactapata Sun Temple. This orientation is clearly demonstrated when sunrise over Machu Picchu is viewed from Llactapata on the morning of the June solstice. Polo de Ondegardo (1965 [1571]) tells us that each Inca village had ceques connecting shrines and it appears possible that a ceque system may have been established at Machu Picchu, smaller but similar to the one in Cusco. An additional ceque possibility in this area is the alignment of Llactapata's Overlook Temple with the River Intihuatana and Cerro Machu Picchu. The potential existence of ceques and huacas other than in Cusco is intriguing. Certain Machu Picchu huaca locations and alignments lend themselves in support of this concept and ultimately may validate Polo de Ondegardo's claim.

The Intihuatana of Machu Picchu in Fig. 9.7 is perhaps the best-known stone of the empire, and certainly at the site. It is finely carved and includes a vertical column in the center of its upper surface. The meaning and purpose of the Intihuatana is subject to debate, but it appears to exhibit no apparent functions regarding any astronomical orientations. It is similar to most of the other intihuatanas in that it is sculpted on all four sides, but the specific carvings of Machu Picchu's Ceremonial Rock bear a much closer resemblance to those found on the River Intihuatana. The stones from the river that were brought to surround the Ceremonial Rock emphasize the potential relationship between it and the one near the river below. The Rumihuasi Stone at Saihuite also is similar and taken together these likenesses suggest there could have been common purpose for the three stones. If the River Intihuatana was indeed an intihuatana, then were the Ceremonial Rock and the Rumihuasi Stone intihuatanas as well? If not, then was the River Intihuatana, first named so by Bingham, really an intihuatana at all? At the bottom of a river valley it certainly was not in a position as favorable to the Sun as were its higher counterparts. The intihuatanas at Machu Picchu, Pisac, Tipon, and the Urubamba River have been described as "hitching posts of the Sun," but perhaps their true functions were somewhat different.

The River Intihuatana, shown in Fig. 9.33, appears that it may have played a more significant role in the overall complex surrounding Machu Picchu than was previously known. It provides a distinct link between the huacas and structures of Machu Picchu, and Llactapata and may be part of a ceque system connecting them both. In addition to lying near the June solstice sunrise/December solstice sunset axis, the River Intihuatana also is part of an orientation between the points of equinox sunrise and sunset that includes Llactapata's Overlook Temple and Cerro Machu Picchu. It is placed at the intersection of the two axes. These potential alignments emphasize the potential significance of the River Intihuatana as it is positioned at this junction. The locations of Llactapata's Sun Temple and Overlook Temple could have been selected specifically to form these alignments. The Sun Temple is remarkable in its orientation with Machu Picchu near to the axis of June solstice sunrise and December solstice sunset, and this is depicted in Fig. 9.48. The rediscovery of Llactapata and all of its many structures has suggested that an intricate system existed throughout the area that was both planned in location and linked in function. The prominence of the River Intihuatana began to emerge as a result of these Llactapata orientations. This appears to be a site of great significance in the development of the overall ceremonial complex.

The location of the River Intihuatana is noteworthy as it lies near the banks of the powerful Urubamba River as well as at the junction of the two solar axes. Upstream, the Urubamba is known as the Vilcanota, the river central to the circulation of cosmological energy from the Earth, through the Milky Way, and then back again by way of sacred mountains. Great power was thought by the Incas to emanate from these waters, thus the river's southeast to northwest orientation makes for a third celestial/cosmological associated alignment that intersects at the site of the River Intihuatana, a correlation of which the Incas could have been well aware.

The features incorporated into this huaca also imply its importance to the Incas. The River Intihuatana exhibits examples of many of the motifs found in other Inca shrines. Its stairs, fountain, basins, and caves, along with the platform and tower, would all indicate that a great effort was put into establishing this sanctuary and of the significance that it held.

The platform at the River Intihuatana is situated such that it has a clear view of both the Llactapata Overlook Temple and the Machu Picchu Intihuatana. Platforms were often constructed for spiritual use and frequently served as ceremonial locations for sacrifices. Such platforms were called *usnu* and Hemming and Ranney (1982, p. 37) tell us that the Spanish chronicler Cabello de Balboa stated "[Huayna Capac] had erected in the square a structure called usnu or chinquin-pillaca, where sacrifices might be offered to the sun and its different phases, with chichi [maize beer] poured out in its honor." The River Intihuatana's platform is adjacent to the Intihuatana, well situated to appease the Sun with appropriate sacrifice.

The second cave in the sanctuary is located beneath the boulder forming the western side of a tower with a rounded wall. The location of this tower at the bottom of a canyon is interesting. Its utility in solar observations would be limited by high horizons, but it still could have been used to note passage of the Sun at zenith. A circular tower reaching to the sky while sitting atop a cave may have served to symbolize and facilitate connections between the world of the present and both the gods of the heavens and the forces of the underworld. The cave could have symbolically provided access to the Pachamama and could also have served to preserve a mummy.

The symbology found at the River Intihuatana appears to have been significant, perhaps diminishing the possibility that it was a simply a stop for pilgrims in the area. Instead the primary function of the huaca might more likely have been to serve as a ceremonial ritual center. The River Intihuatana may have been a key component in the power and significance of the greater Machu Picchu complex.

The exploration of Llactapata has given rise to many new questions regarding the overall extent, orientation, and function of the entire Machu Picchu complex. Its many structures suggest a significant enlargement of the overall estate and support the possibility that there may have been a ceque system here as well as the one located in Cusco. The orientation of the stone-lined channel and Sun Temple at Llactapata with the River Intihuatana and Machu Picchu's Sacred Plaza is striking and, as shown in Figs. 9.48 and 9.49, can be clearly seen from Llactapata at sunrise on the day of the June solstice. Orientations with the Pleiades exist both from Llactapata's Sun Temple and at Machu Picchu's Torreon.

Another interesting hypothesis concerns a role that Llactapata may have played in the ceremonies of Machu Picchu. The two sites appear to be directly associated with one another and it becomes likely that Llactapata was an extension of and supported ritual activities in the area. The demonstration shown in Fig. 9.51 with solar reflections from Llactapata's Sun Temple to Machu Picchu's Sacred Plaza at the time of the June solstice indicates that solar ceremonies including the two sites were possible.

10.2 Findings

The previous chapters highlighted 31 sites in and surrounding Cusco, in the Sacred Valley, and in the Machu Picchu region. 24 of these sites are either carved rock huacas or sanctuaries that include carved or otherwise significant rocks. The remaining seven are huacas or sites with structures that lack intrinsic rock. In this section these rock and non-rock huacas/sites are categorized as to whether or not they exhibit astronomical orientation. The Cusco region includes sites at 19 locations, nine of the sites are in the Sacred Valley, and the remaining three are related to Machu Picchu.

Chapter 1 introduced four main points of interest for examination. They are revisited here with findings derived from data discussed in Chaps. 7, 8, and 9.

The first main point examined was:

> *The majority of the currently identified rock huacas in the greater Cusco Valley region, the Sacred Valley, and surrounding Machu Picchu are associated with visual solar phenomena.*

The sites are divided into two groups – those that were or included rock shrines and those that didn't (Tables 10.1, 10.2, and 10.3). For the purposes of these tables, a huaca was considered astronomical if it, or an element of it, was found to have a solar light and shadow effect or alignment relating in any way to the Sun. The book's primary focus was to examine the solar culture of the Incas. Other interests, such as in the Pleiades, were presented in earlier chapters to give a much broader sense of Inca astronomy as a whole but are not included here.

Of the 31 sites examined, 25, or 81%, were found to fit at least one of the solar criteria. 16 of the sites included sacred rocks with celestial orientations and nine more had structures oriented with an aspect of the solar horizon. This equates to 52% and 29% respectively. The 16 sites that include rocks with alignments are categorized as "Astronomical Rock." Sites such as Sacsahuaman, Cerro Unoraqui, and the River Intihuatana have sacred rocks, but those rocks were not found to be specifically encoded with astronomy. These sites did exhibit other astronomical orientations, however, and were included with the ones displaying structural alignments in the "Astronomical Non-Rock" category. Four huacas with boulders

Table 10.1 Classifications for the Cusco Region, Tipon, and Saihuite (author's table)

Section	Site	Astronomical Rock	Astronomical Non-Rock	Non-Astronomical Rock	Non-Astronomical Non-Rock
7.1	Kenko Grande	X			
7.2	Kenko Chico	X			
7.3	Mesa Redonda			X	
7.4	Tetecaca			X	
7.5	Patallacta			X	
7.6	Kusilluchayoc	X			
7.7	Lacco	X			
7.8	Huaca 44	X			
7.9	Lanlakuyok	X			
7.10	Puca Pucara		X		
7.11	Tambomachay	X			
7.12	Sacsahuaman		X		
7.13	Mollanguanca	X			
7.14	Sapatiana			X	
7.15	Rumiwasi Bajo		X		
7.16	Rumiwasi Alto		X		
7.17	Kusicallanca		X		
7.18	Tipon	X			
7.19	Saihuite	X			

Table 10.2 Classifications for the Sacred Valley (author's table)

Section	Site	Astronomical Rock	Astronomical Non-Rock	Non-Astronomical Rock	Non-Astronomical Non-Rock
8.1	Chinchero	X			
8.2	Pisac	X			
8.3	Q'espiwanka	X			
8.4	Cerro Pumahuachana				X
8.5	Cerro Unoraqui		X		
8.6	Maras				X
8.7	Moray		X		
8.8	Choquequilla	X			
8.9	Ollantaytambo	X			

showed no evidence of solar alignments and two of the sites without a rock had none. Non-astronomical huacas comprised 19% of the sites that were examined. The "Non-Astronomical Rock" huacas (without observable solar orientations) were Mesa Redonda, Tetecaca, Patallacta, and Sapantiana, each among the closest

Table 10.3 Classifications for the Machu Picchu Region (author's table)

Section	Site	Astronomical Rock	Astronomical Non-Rock	Non-Astronomical Rock	Non-Astronomical Non-Rock
9.2	Machu Picchu	X			
9.3	River Intihuatana		X		
9.4	Llactapata		X		

to Cusco. They likely served purposes independent of the Sun. The other sites without specific solar alignment ("Non-Astronomical Non-Rock") were the ridgeline of Cerro Pumahuachana and the Moras salt mines. The existence of orientations with the Sun at 81% of the sites supports this first main point of interest and implies that additional Inca huacas may also have been associated with solar observation and ceremony. Percentage summaries are given in Tables 10.16, 10.17, 10.18, and 10.19.

Sites in the following tables are ordered by their chapter section numbers.

The second main point for research was:

Those huacas found to be associated with visual solar phenomena (solar huacas) exhibit alignments related to the solstices, equinoxes, zenith Suns and anti-zenith Suns.

Orientations for the solstices were found at 84% (21) of the solar huacas and 40% (10) displayed specific effects of light and shadow. 32% (8) had orientations for the equinoxes and 24% (6) had zenith or anti-zenith alignments (Tables 10.4, 10.5 and 10.6).

63% (10) of the 16 astronomical rock huacas had orientations for the June solstice sunrise, 38% (6) for the June solstice sunset, 56% (9) for the December solstice sunrise and 25% (4) for the December solstice sunset. 31% (5) had orientations for an equinox sunrise and 25% (4) for the equinox sunset. There were four instances of zenith Sun alignments (25%) and one (6%) anti-zenith orientation.

The number of examples for June and December solstice is fairly even, somewhat unexpected since the solstice in December is in the rainy season when observations of the horizon on specific days would be expected to be far less reliable. There is a marked difference between the numbers of alignments for solstice sunrises in both seasons when compared to those for the associated sunsets. Results imply a greater ceremonial interest in the rising Sun.

Three of the nine astronomical non-rock huacas had June solstice sunrise alignments (33%), none for the June solstice sunset, three for the December solstice sunrise (33%) and none for the December solstice sunset.

Of the 25 sites with any solar association, 96% held at least one orientation for one of the six primary horizon events of sunrise and sunset at June solstice, December solstice, and the equinoxes. There were a total of seven possible examples of zenith Sun alignments and two instances of potential alignments related to the anti-zenith Sun. While there were fewer examples regarding the zenith Sun and

Table 10.4 Solar Orientations Identified at Cusco, Tipon, and Saihuite (author's table)

Section	Site	JSSR	JSSS	DSSR	DSSS	ESR	ESS	ZSR	AZSR
7.1	Kenko Grande	X		X		X	X		
7.2	Kenko Chico			X					
7.3	Mesa Redonda								
7.4	Tetecaca								
7.5	Patallacta								
7.6	Kusilluchayoc		X						
7.7	Lacco	X						X	
7.8	Huaca 44	X	X	X	X	X	X	X	
7.9	Lanlakuyok					X			
7.10	Puca Pucara					X	X		
7.11	Tambomachay			X					
7.12	Sacsahuaman							X	X
7.13	Mollaguanca	X			X				
7.14	Sapatiana								
7.15	Rumiwasi Bajo			X					
7.16	Rumiwasi Alto			X					
7.17	Kusicallanca			X					
7.18	Tipon		X						
7.19	Saihuite	X							

JSSR June Solstice Sun Rise, *DSSR* December Solstice Sun Rise, *ESR* Equinox Sun Rise, *JSSS* June Solstice Sun Set, *DSSS* December Solstice Sun Set, *ESS* Equinox Sun Set, *ZSR* Zenith Sun Rise, *AZSR* Anti-Zenith Sun Rise

Table 10.5 Solar Orientations Identified in the Sacred Valley (author's table)

Section	Site	JSSR	JSSS	DSSR	DSSS	ESR	ESS	ZSR	AZSR
8.1	Chinchero	X			X	X	X		
8.2	Pisac	X		X			X		
8.3	Q'espiwanka	X	X	X					
8.4	Cerro Pumahuachana								
8.5	Cerro Unoraqui	X							
8.6	Maras								
8.7	Moray	X						X	
8.8	Choquequilla			X					
8.9	Ollantaytambo	X	X	X				X	

JSSR June Solstice Sun Rise, *DSSR* December Solstice Sun Rise, *ESR* Equinox Sun Rise, *JSSS* June Solstice Sun Set, *DSSS* December Solstice Sun Set, *ESS* Equinox Sun Set, *ZSR* Zenith Sun Rise, *AZSR* Anti-Zenith Sun Rise

anti-zenith Sun with these rural huacas, those events appeared to be of greater interest at Cusco. June solstice sunrise orientations were noted 20 times (including multiple alignments at the same site), June solstice sunset seven times, December solstice sunrise 12 times, December solstice sunset eight times, equinox sunrise 10 times and equinox sunset nine times. While major sites such as Kenko Grande,

Table 10.6 Solar Orientations Identified in the Machu Picchu Region (author's table)

Section	Site	JSSR	JSSS	DSSR	DSSS	ESR	ESS	ZSR	AZSR
9.2	Machu Picchu	X	X	X	X	X		X	X
9.3	River Intihuatana					X		X	
9.4	Llactapata	X							

JSSR June Solstice Sun Rise, *DSSR* December Solstice Sun Rise, *ESR* Equinox Sun Rise, *JSSS* June Solstice Sun Set, *DSSS* December Solstice Sun Set, *ESS* Equinox Sun Set, *ZSR* Zenith Sun Rise, *AZSR* Anti-Zenith Sun Rise

Table 10.7 Water Sources Found with Sites at Cusco, Tipon, and Saihuite (author's table)

Section	Site	Water Source
7.1	Kenko Grande	X
7.2	Kenko Chico	X
7.3	Mesa Redonda	X
7.4	Tetecaca	X
7.5	Patallacta	
7.6	Kusilluchayoc	X
7.7	Lacco	X
7.8	Huaca 44	
7.9	Lanlakuyok	X
7.10	Puca Pucara	X
7.11	Tambomachay	X
7.12	Sacsahuaman	X
7.13	Mollaguanca	X
7.14	Sapatiana	X
7.15	Rumiwasi Bajo	
7.16	Rumiwasi Alto	
7.17	Kusicallanca	
7.18	Tipon	X
7.19	Saihuite	X

Saihuite, Chinchero, and Machu Picchu were found to have multiple solar horizon orientations, many of the rest exhibited only one or two.

The second main point was supported in part as 96% of the solar huacas appear to have incorporated direct observation of the solstices and equinoxes.

The third main point explored was:

Solar huacas are associated with flowing water.

Water sources were found to be associated with 72% (21) of the huacas examined. With the exception of Huaca 44 (Sect. 7.8), the 16 astronomical rock huacas all had sources of water (94%). While no evidence was found for water at this site, due to its proximity to Lacco and the Ucu Ucu canal, it cannot be ruled out that water was once provided to this shrine as well (Tables 10.7, 10.8, and 10.9). Therefore, it is possible to suggest that 100% of the astronomical rock huacas in this study may have had sources of water for camay (Sect. 3.4) and other purposes.

Table 10.8 Water Sources Found with Sites in the Sacred Valley (author's table)

Section	Site	Water Source
8.1	Chinchero	X
8.2	Pisac	X
8.3	Q'espiwanka	X
8.4	Cerro Pumahuachana	
8.5	Cerro Unoraqui	
8.6	Maras	X
8.7	Moray	
8.8	Choquequilla	X
8.9	Ollantaytambo	X

Table 10.9 Water Sources Found with Sites in the Machu Picchu Region (author's table)

Section	Site	Water Source
9.2	Machu Picchu	X
9.3	River Intihuatana	X
9.4	Llactapata	X

Of the non-astronomical rock huacas near Cusco, only Patallacta exhibited no evidence of a water supply, but there are canals in the area and modern construction of roads and buildings may mask the path of ancient hydraulics. Immediate sources of water were not identified at the astronomical non-rock sites of Rumiwasi Bajo, Rumiwasi Alto, and Kusicallanca. The peak of Cerro Unoraqui was understandably without a water source, as was the Sun Temple on the Llactapata ridge. Water would have had to have been carried to the channel that leads from the Sun Temple toward Machu Picchu and the June solstice sunrise.

100% (16) of the solar rock huacas were found to have or were likely to have had sources of flowing water. Only 33% (3) of the nine astronomical non-rock sites were found to have water sources, either present or past. The third main point was supported for huacas made of rock, but not for other huacas. This finding is in keeping with forces of camay being employed to bring life to these venerated stones.

The fourth main point of interest was:

Solar huacas in the form of carved rocks contain traditional Andean motifs.

This again was a point with results differing for rock huacas and non-rock huacas. 56% (9) of the 16 astronomical rock huacas had shamanic stairs, 38% (6) had caves, 69% (11) employed niches, 75% (12) exhibited seats, 31% (5) had basins, 38% (6) had animal carvings, 31% (5) had channels for ceremonial fluids, and 38% (6) had fountains. 100% of the astronomical rock huacas exhibited at least one or more of these motifs (Tables 10.10, 10.11, and 10.12).

Of the nine astronomical sites without celestially oriented rocks, 22% (2) had stairs, 33% (3) were associated with caves, 67% (6) had niches, 11% (1) had seats, 22% (2) displayed basins, none had animal carvings, 11% (1) had a ceremonial channel and 22% (2) had fountains.

Table 10.10 Andean Motifs Found with Sites at Cusco, Tipon, and Saihuite (author's table)

Section	Site	Shamanic Stairs	Caves	Niches	Seats	Basins	Animal Carvings	Ceremonial Channels	Fountains
7.1	Kenko Grande	X	X	X	X	X	X	X	
7.2	Kenko Chico	X			X	X			
7.3	Mesa Redonda	X							
7.4	Tetecaca		X	X				X	
7.5	Patallacta				X				
7.6	Kusilluchayoc			X			X	X	
7.7	Lacco	X	X	X	X	X	X	X	
7.8	Huaca 44				X				
7.9	Lanlakuyok	X	X	X	X				
7.10	Puca Pucara								X
7.11	Tambomachay								X
7.12	Sacsahuaman	X	X	X	X	X			
7.13	Mollaguanca				X				
7.14	Sapatiana						X		
7.15	Rumiwasi Bajo		X	X					
7.16	Rumiwasi Alto			X					
7.17	Kusicallanca			X					
7.18	Tipon			X					X
7.19	Saihuite	X		X	X		X	X	X

Table 10.11 Andean Motifs Found with Sites in the Sacred Valley (author's table)

Section	Site	Shamanic Stairs	Caves	Niches	Seats	Basins	Animal Carvings	Ceremonial Channels	Fountains
8.1	Chinchero	X	X	X	X	X	X	X	
8.2	Pisac			X					X
8.3	Q'espiwanka			X					
8.4	Cerro Pumahuachana								
8.5	Cerro Unoraqui								
8.6	Maras								
8.7	Moray								
8.8	Choquequilla		X	X					
8.9	Ollantaytambo				X				X

As might be expected, major sites such as Kenko Grande, Lacco, Saihuite, Chinchero, and Machu Picchu exhibited the greatest number of these motifs. Most of the huacas without astronomical rocks displayed only one or two. This main point was largely supported for major carved rock solar huacas, but not as much for the others.

Table 10.12 Andean Motifs Found with Sites in the Machu Picchu Region (author's table)

Section	Site	Shamanic Stairs	Caves	Niches	Seats	Basins	Animal Carvings	Ceremonial Channels	Fountains
9.2	Machu Picchu	X	X	X	X	X	X		X
9.3	River Intihuatana	X	X	X		X			X
9.4	Llactapata			X				X	

Table 10.13 Huaca Features in the Cusco Region (author's table)

Section	Site	Sacred Rock	Light and Shadow	Cardinal Alignments	Light-Tubes	Altars	Platforms	Gnomons or Pillars	Horizon Replicas
7.1	Kenko Grande	X	X		X	X		X	
7.2	Kenko Chico	X						X	
7.3	Mesa Redonda	X				X			
7.4	Tetecaca	X				X		X	
7.5	Patallacta	X							
7.6	Kusilluchayoc	X							
7.7	Lacco	X	X		X	X		X	
7.8	Huaca 44	X						X	
7.9	Lanlakuyok	X			X				
7.10	Puca Pucara			X			X		
7.11	Tambomachay	X					X		
7.12	Sacsahuaman	X				X			
7.13	Mollaguanca	X							
7.14	Sapatiana	X							
7.15	Rumiwasi Bajo								
7.16	Rumiwasi Alto								
7.17	Kusicallanca								
7.18	Tipon	X							
7.19	Saihuite	X	X	X			X		

Additional Motifs and Features

The remaining motifs and features recorded at the sites are categorized in Tables 10.13, 10.14, and 10.15. Effects of light and shadow were identified at 31% (9) of the sites and cardinal alignments were found at 17% (5). Light-tubes were used in 14% (4) of the locations and altars were employed at 24% (7). There were platforms at 21% (6) of the sites and 34% (10) had gnomons or pillars. As with the features listed in Tables 10.10, 10.11, and 10.12, the primary shrines and sanctuaries included more of these elements than did the others.

Table 10.14 Huaca Features in the Sacred Valley Region (author's table)

Section	Site	Sacred Rock	Light and Shadow	Cardinal Alignments	Light-Tubes	Altars	Plat-forms	Gnomons or Pillars	Horizon Replicas
8.1	Chinchero	X	X	X		X			
8.2	Pisac	X	X				X		
8.3	Q'espiwanka	X						X	
8.4	Cerro Pumahuachana								
8.5	Cerro Unoraqui	X		X				X	
8.6	Maras								
8.7	Moray		X						
8.8	Choquequilla	X							
8.9	Ollantaytambo	X	X					X	

Table 10.15 Huaca Features in the Machu Picchu Region (author's table)

Section	Site	Sacred Rock	Light and Shadow	Cardinal Alignments	Light-Tubes	Altars	Platforms	Gnomons or Pillars	Horizon Replicas
9.2	Machu Picchu	X	X		X	X	X	X	X
9.3	River Intihuatana	X	X	X			X	X	
9.4	Llactapata		X						

Table 10.16 Percentages of Astronomical Alignments per Category (author's chart)

Category	Horizon Events	JSSR	JSSS	DSSR	DSSS	ESR	ESS	ZSR	AZSR
Astronomical Rock	100%	63%	38%	56%	25%	31%	25%	25%	6%
Astronomical Non-Rock	100%	33%	0%	38%	0%	25%	13%	33%	13%

JSSR June Solstice Sunrise, *DSSR* December Solstice Sunrise, *ESR* Equinox Sunrise, *JSSS* June Solstice Sunset, *DSS* December Solstice Sunset, *ESS* Equinox Sunset, *ZSR* Zenith Sunrise, *AZSR* Anti-Zenith Sunrise, *Ast Rock* Astronomical Rock, *Ast NRock* Astronomical Non-Rock, *NAst Rock* Non-Astronomical Rock, *Nast NRock* Non-Astronomical Non-Rock

Statistical Summaries

Percentages for each recorded feature per huaca category are given in Tables 10.16, 10.17, 10.18, and 10.19. Astronomical Rock Huacas led or tied in all but one of the categories in Table 10.16 when compared with the Astronomical Non-Rock category. By definition the non-astronomical categories are not represented. Orientations for sunrises on the June and December solstices were the most common and, also by definition, 100% of the Astronomical Rock and Astronomical Non-Rock huacas

Table 10.17 Percentage per Category with Water Sources (author's table)

Category	Water Source
Astronomical Rock	100%
Astronomical Non-Rock	44%
Non-Astronomical Rock	75%
Non-Astronomical Non-Rock	50%

Table 10.18 Percentages of Andean Motifs per Category (author's table)

Category	Shamanic Stairs	Caves	Niches	Seats	Basins	Animal Carvings	Ceremonial Channels	Fountains
Astronomical Rock	56%	38%	69%	75%	31%	38%	31%	38%
Astronomical Non-Rock	22%	33%	56%	11%	22%	0%	11%	22%
Non-Astronomical Rock	25%	25%	25%	25%	0%	25%	25%	0%
Non-Astronomical Non-Rock	0%	0%	0%	0%	0%	0%	0%	0%

Table 10.19 Percentages of Features per Huaca Category (author's table)

Category	Sacred Rock	Light and Shadow	Cardinal Alignments	Light-Tubes	Altars	Platforms	Gnomons or Pillars	Horizon Replicas
Astronomical Rock	100%	38%	13%	25%	25%	25%	44%	6%
Astronomical Non-Rock	33%	33%	33%	0%	11%	11%	22%	0%
Non-Astronomical Rock	100%	0%	0%	0%	50%	0%	25%	0%
Non-Astronomical Non-Rock	0%	0%	0%	0%	0%	0%	0%	0%

were found to have alignments for "horizon events." The remaining columns of Table 10.16 give a breakdown for each individual solar event. An associated graph is presented as Fig. 10.2.

As described in Table 10.17, all rock huacas were found likely to have been associated with water, while less than half of the non-rock sites exhibited such hydraulic sources. Table 10.18 shows that seats, niches, and stairs were the most common motifs found on or near the astronomical rock huacas and niches built into walls were the predominant feature in the astronomical non-rock huacas. An associated graph is presented as Fig. 10.3. Table 10.19 follows with cardinal alignments having

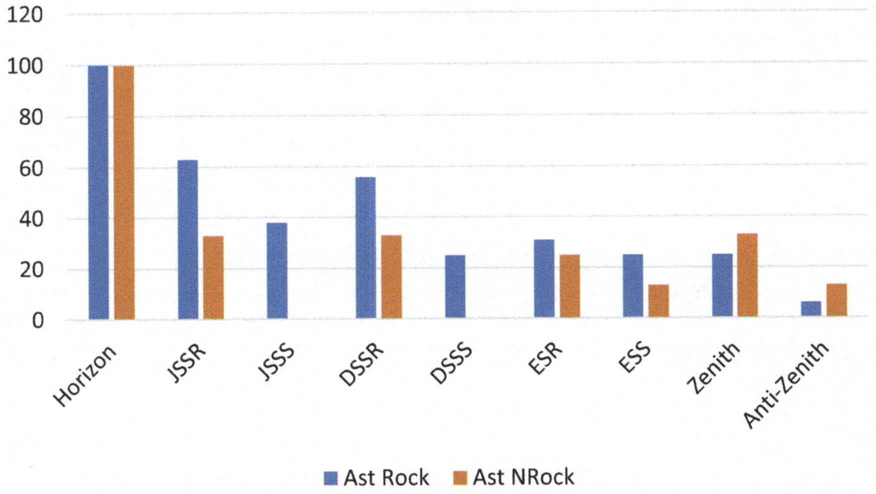

Fig. 10.2 Percentages of Astronomical Alignments per Category (author's graph). JSSR June Solstice Sunrise, DSSR December Solstice Sunrise, ESR Equinox Sunrise, JSSS June Solstice Sunset, DSS December Solstice Sunset, ESS Equinox Sunset, ZSR Zenith Sunrise, AZSR Anti-Zenith Sunrise, Ast Rock Astronomical Rock, Ast NRock Astronomical Non-Rock, NAst Rock Non-Astronomical Rock, Nast NRock Non-Astronomical Non-Rock

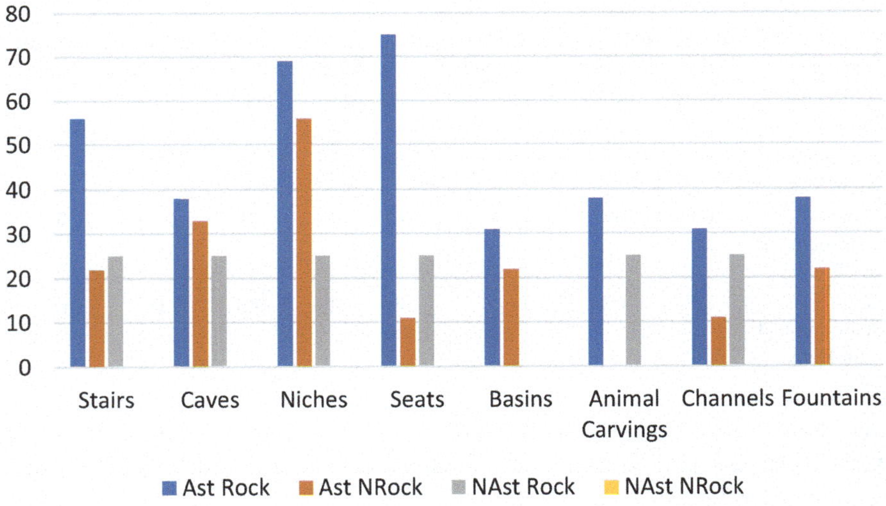

Fig. 10.3 Percentages of Andean motifs per category (author's graph). JSSR June Solstice Sunrise, DSSR December Solstice Sunrise, ESR Equinox Sunrise, JSSS June Solstice Sunset, DSS December Solstice Sunset, ESS Equinox Sunset, ZSR Zenith Sunrise, AZSR Anti-Zenith Sunrise, Ast Rock Astronomical Rock, Ast NRock Astronomical Non-Rock, NAst Rock Non-Astronomical Rock, Nast NRock Non-Astronomical Non-Rock

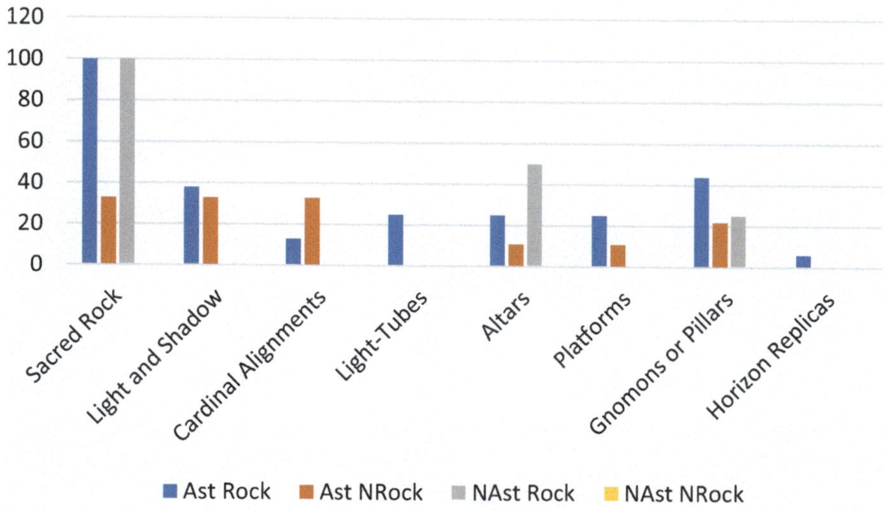

Fig. 10.4 Percentages of features per huaca category (author's graph). JSSR June Solstice Sunrise, DSSR December Solstice Sunrise, ESR Equinox Sunrise, JSSS June Solstice Sunset, DSS December Solstice Sunset, ESS Equinox Sunset, ZSR Zenith Sunrise, AZSR Anti-Zenith Sunrise, Ast Rock Astronomical Rock, Ast NRock Astronomical Non-Rock, NAst Rock Non-Astronomical Rock, Nast NRock Non-Astronomical Non-Rock

been found in the walls of 33% of astronomical non-rock huacas and gnomons or pillars associated with 44% of astronomical rock huacas. An associated graph is presented as Fig. 10.4.

10.3 Concluding Remarks

The Incas possessed sophisticated celestial knowledge, and as solar worshippers they chose to incorporate and display alignments and features of the Sun, their god, in their many temples and shrines across the empire. A picture emerges of a culture interwoven with an astronomy, which was employed as a tool to guide many aspects of daily life. The huacas reveal a society that was infatuated with the Sun and that possessed the technical ability to accurately encode its celestial knowledge into structures and carvings.

The intent of this book is to provide to others the information, images, and data that I gathered in Peru. It is my hope that what was presented throughout may add insight useful in further research. Appendices A.2 through A.5 provide additional data to help in that regard.

Appendices

A.1 Glossary

Aclla – a chosen woman serving the ruling Inca
Acllahuasi – house of the chosen women
Amaru – snake
Anti – dawn or east
Antisuyu – the northeast quadrant of both Cusco and of the Inca Empire
Apu – sacred mountain
Ayllu – a non-royal kin group
Camayoc – attendant or specialist
Cancha – an enclosed compound and courtyard
Capac – royal
Capac Raymi – a festival at the time of the December solstice
Ceque – a sightline that organizes huacas around its direction
Chaca – bridge
Chasqui – courier
Chicha – maize beer
Chinchaysuyu – the northwest quadrant of both Cusco and of the Inca Empire
Chulpa – burial tower
Citua Raymi – Inca festival in September
Cocha – lake
Collana – something that is principal or first
Collasuyu – the southeast quadrant of both Cusco and of the Inca Empire
Coricancha – the principal Sun temple of Cusco
Coya – the sister-wife of a ruling Inca
Coya Raymi – see Citua Raymi
Cunti – sunset, the west
Cuy – guinea pig
Cuntisuyu – the southwest quadrant of both Cusco and of the Inca Empire

© Springer Nature Switzerland AG 2020

S. R. Gullberg, *Astronomy of the Inca Empire*, Historical & Cultural Astronomy,
https://doi.org/10.1007/978-3-030-48366-1

Fig. A.1 Women with street food (watercolor by Jessica Gullberg)

Curaca – chief or leader
Hanan – upper
Hatun – large or great
Huaca – shrine
Huaman – falcon
Huayna – young
Huasi – house
Huatana – hitching post
Huauque – totem
Huchuy – small
Hurin – lower
Illapa – thunder
Inti – the Sun
Intihuatana – hitching post of the Sun
Inti Raymi – a festival at the time of the June solstice
Kallanka – long rectangular hall with many doors
Kenko – zigzagged
Llacta – town

Machay – cave
Mama – mother
Mamacona – holy woman
Mayu – river
Mita – labor service for the Empire
Mitimae – resettled colonists
Moiety – One of two parts. Inca communities were often divided into upper and lower sectors
Ñusta – princess
Pacarina – place of origin of mythical ancestors as they emerged from the Earth
Pacariqtambo – place of Inca origin
Paccha – waterfall
Pacha – Earth
Pampa – plain
Panaca – a kin group formed of the royal descendants of an Inca
Pata – platform or terrace
Picchu – mountain
Pirua – storehouse
Puma – mountain lion
Puna – high savannah
Punchao – a golden Sun idol
Puncu – entrance or gate
Puquiu – a spring or fountain
Pururaucas – stones mythically transformed into warriors when the Inca defeated the Chanca
Qollca – storehouse
Quilla – the Moon, also a month
Quipu – a knotted cord used to record information
Quipucamayoc – person who kept, recorded and interpreted quipus
Racay – hall or shed
Raymi – festival
Rimac – oracle
Rumi – stone
Sayhua – marker
Sucanca – solar marker
Suyu – a geographical division. The region of Cusco was divided into four quadrants, or suyus
Tahuantinsuyu – the name for the Inca Empire as a whole
Tambo – place of lodging
Taqui – dance
Tocco – a cave
Urco – hill
Usnu – administrative or religious platform
Vilca – sacred
Yunga – lowland

Wak'a – huaca, shrine
Warachikoy – a December ritual for recognizing young noblemen as adults
Zeq'e – ceque, line

A.2 Azimuths and Inclinations

Cusco Region, Tipon, and Saihuite

7.1 Kenko Grande

Rear Altar

Calculated Equinox Sunrise Azimuth:	089.9°
Measured Azimuth:	094.0°
Magnetic Declination:	3.9° W
Corrected Azimuth:	090.1°
Difference Calculated vs. Corrected:	0.2°
Measured Inclination:	+2°

Fig. A.2 Rumihuasi Stone at Saihuite (pencil by Jessica Gullberg)

7.2 Kenko Chico

Calculated December Solstice Sunrise Azimuth:	113.6°
Measured Azimuth:	116.0°
Magnetic Declination:	3.9° W
Corrected Azimuth:	112.1°
Difference Calculated vs. Corrected:	1.5°
Measured Inclination:	+2°

7.6 Kusilluchayoc

Calculated June Solstice Sunrise Azimuth:	295.3°
Measured Azimuth:	297.5°
Magnetic Declination:	3.9° W
Corrected Azimuth:	293.6°
Difference Calculated vs. Corrected:	1.7°
Measured Inclination:	+4°

7.7 Lacco

Northeast Cave

Calculated June Solstice Sunrise Azimuth:	065.0°
Measured Azimuth:	067.5°
Magnetic Declination:	3.9° W
Corrected Azimuth:	063.6°
Difference Calculated vs. Corrected:	1.4°
Measured Inclination:	+3°

7.8 Huaca 44

Calculated June Solstice Sun Rise Azimuth:	065.0°
Measured Azimuth:	068.0°
Magnetic Declination:	3.9° W
Corrected Azimuth:	064.1°
Difference Calculated vs. Corrected:	0.9°
Measured Inclination:	+3°

Calculated Equinox Sun Rise Azimuth:	089.6°
Measured Azimuth:	092.5°

Magnetic Declination:	3.9° W
Corrected Azimuth:	088.6°
Difference Calculated vs. Corrected:	1.0°
Measured Inclination:	+2°

Calculated December Solstice Sun Rise Azimuth:	113.6°
Measured Azimuth:	115.0°
Magnetic Declination:	3.9° W
Corrected Azimuth:	111.1°
Difference Calculated vs. Corrected:	2.5°
Measured Inclination:	+2°

7.9 Lanlakuyok

Calculated Equinox Sun Rise Azimuth:	089.6°
Measured Azimuth:	091.5°
Magnetic Declination:	3.9° W
Corrected Azimuth:	087.6°
Difference Calculated vs. Corrected:	2.0°
Measured Inclination:	+2°

7.10 Puca Pucara

Calculated Equinox Sun Rise Azimuth:	089.6°
Measured Azimuth:	093.0°
Magnetic Declination:	3.9° W
Corrected Azimuth:	089.1°
Difference Calculated vs. Corrected:	0.5°
Measured Inclination:	+2°

Calculated Equinox Sun Set Azimuth:	270.5°
Measured Azimuth:	273.0°
Magnetic Declination:	3.9° W
Corrected Azimuth:	269.6°
Difference Calculated vs. Corrected:	0.4°
Measured Inclination:	+1°

7.11 Tambomachay

Calculated June Solstice Sun Rise Azimuth:	113.9°
Measured Azimuth:	118.0°

Magnetic Declination:	3.9° W
Corrected Azimuth:	114.1°
Difference Calculated vs. Corrected:	0.2°
Measured Inclination:	+1°

7.13 Mollaguanca

Calculated June Solstice Sun Rise Azimuth:	065.0°
Measured Azimuth:	068.0°
Magnetic Declination:	3.9° W
Corrected Azimuth:	064.1°
Difference Calculated vs. Corrected:	0.9°
Measured Inclination:	+3°

Calculated December Solstice Sun Set Azimuth:	246.4°
Measured Azimuth:	248.0°
Magnetic Declination:	3.9° W
Corrected Azimuth:	245.1°
Difference Calculated vs. Corrected:	1.3°
Measured Inclination:	+4°

7.15 Rumiwasi Bajo

Calculated December Solstice Sun Rise Azimuth:	113.9°
Measured Azimuth:	116.0°
Magnetic Declination:	3.9° W
Corrected Azimuth:	112.1°
Difference Calculated vs. Corrected:	1.8°
Measured Inclination:	+2°

Calculated June Solstice Sun Set Azimuth:	294.4°
Measured Azimuth:	286.0°
Magnetic Declination:	3.9° W
Corrected Azimuth:	282.1°
Difference Calculated vs. Corrected:	12.3°
Measured Inclination:	+1°

7.16 Rumiwasi Alto

Calculated December Solstice Sun Rise Azimuth:	113.9°
Measured Azimuth:	116.5°

Magnetic Declination:	3.9° W
Corrected Azimuth:	112.6°
Difference Calculated vs. Corrected:	1.3°
Measured Inclination:	+1°

7.17 Kusicallanca

Calculated December Solstice Sun Rise Azimuth:	114.1°
Measured Azimuth:	120.0°
Magnetic Declination:	3.9° W
Corrected Azimuth:	116.1°
Difference Calculated vs. Corrected:	2.0°
Measured Inclination:	0°

7.18 Tipon

Iglesia Raqui and Northeastern Structure

Calculated June Solstice Sun Set Azimuth:	298.4°
Measured Azimuth:	301.5°
Magnetic Declination:	4.1° W
Corrected Azimuth:	297.4°
Difference Calculated vs. Corrected:	1.0°
Measured Inclination:	+13°

Intihuatana

Calculated June Solstice Sun Set Azimuth:	294.4°
Measured Azimuth:	298.0°
Magnetic Declination:	4.1° W
Corrected Azimuth:	293.9°
Difference Calculated vs. Corrected:	0.5°
Measured Inclination:	+1°

7.19 Saihuite Upper Structure

Calculated June Solstice Sun Rise Azimuth:	065.3°
Measured Azimuth:	067.5°
Magnetic Declination:	3.2°W
Corrected Azimuth:	064.3°

Difference Calculated vs. Corrected:	1.0°
Measured Inclination:	+2°

The Sacred Valley

8.1 Chinchero

Chinkana toward Titikaka

Calculated December Solstice Sun Set Azimuth:	248.1°
Measured Azimuth:	248.0°
Magnetic Declination:	3.7° W
Corrected Azimuth:	244.3°
Difference Calculated vs. Corrected:	3.8°
Measured Inclination:	+10°

Titikaka toward Chinkana

Calculated June Solstice Sun Rise Azimuth:	064.5°
Measured Azimuth:	068.0°
Magnetic Declination:	3.7° W
Corrected Azimuth:	064.3°
Difference Calculated vs. Corrected:	0.2°
Measured Inclination:	+5°

8.2 Pisac

Intihuatana

Calculated June Solstice Sun Rise Azimuth:	065.9°
Measured Azimuth:	068.0°
Magnetic Declination:	4.0° W
Corrected Azimuth:	064.0°
Difference Calculated vs. Corrected:	1.9°
Measured Inclination:	0°

Platform

Calculated December Solstice Sun Rise Azimuth:	114.1°
Measured Azimuth:	116.0°

Magnetic Declination:	4.0° W
Corrected Azimuth:	112.0°
Difference Calculated vs. Corrected:	2.1°
Measured Inclination:	0°

8.3 Q'espiwanka

Boulder toward Solar Pillars

Calculated June Solstice Sun Rise Azimuth:	058.4°
Measured Azimuth:	061.0°
Magnetic Declination:	3.8° W
Corrected Azimuth:	057.2°
Difference Calculated vs. Corrected:	1.2°
Measured Inclination:	+22°

Outside South Wall

Calculated June Solstice Sun Rise Azimuth:	058.9°
Measured Azimuth:	061.5°
Magnetic Declination:	3.8° W
Corrected Azimuth:	057.7°
Difference Calculated vs. Corrected:	1.2°
Measured Inclination:	+21°

Boulder toward Cerro Unoraqui

Calculated December Solstice Sun Rise Azimuth:	112.7°
Measured Azimuth:	114.0°
Magnetic Declination:	3.6° W
Corrected Azimuth:	110.4°
Difference Calculated vs. Corrected:	2.3°
Measured Inclination:	+6°

Boulder toward other Solar Horizon Events

Calculated June Solstice Sun Set Azimuth:	295.5°
Measured Azimuth:	298.0°
Magnetic Declination:	3.6° W
Corrected Azimuth:	294.4°
Difference Calculated vs. Corrected:	1.1°
Measured Inclination:	+5°

Calculated December Solstice Sun Set Azimuth:	247.1°
Measured Azimuth:	061.0°
Magnetic Declination:	3.6° W
Corrected Azimuth:	057.2°
Difference Calculated vs. Corrected:	1.2°
Measured Inclination:	+5°

Calculated Equinox Sun Rise Azimuth:	087.0°
Measured Azimuth:	090.0°
Magnetic Declination:	3.6° W
Corrected Azimuth:	086.4°
Difference Calculated vs. Corrected:	0.6°
Measured Inclination:	+14°

Calculated Equinox Sun Set Azimuth:	270.8°
Measured Azimuth:	273.0°
Magnetic Declination:	3.6° W
Corrected Azimuth:	269.4°
Difference Calculated vs. Corrected:	1.4°
Measured Inclination:	+5°

Calculated Zenith Sun Rise Azimuth:	103.1°
Measured Azimuth:	106.0°
Magnetic Declination:	3.6° W
Corrected Azimuth:	102.4°
Difference Calculated vs. Corrected:	0.7°
Measured Inclination:	+4°

Calculated Anti Zenith Sun Set Azimuth:	284.9°
Measured Azimuth:	283.4°
Magnetic Declination:	3.6° W
Corrected Azimuth:	057.2°
Difference Calculated vs. Corrected:	1.5°
Measured Inclination:	+5°

8.8 Choquequilla

Calculated December Solstice Sun Rise Azimuth:	109.5°
Measured Azimuth:	114.0°
Magnetic Declination:	3.6° W
Corrected Azimuth:	100.4°
Difference Calculated vs. Corrected:	0.9°
Measured Inclination:	+32°

Machu Picchu Region

9.2 Machu Picchu

Intihuatana

Calculated June Solstice Sun Rise Azimuth:	061.8°
Measured Azimuth:	064.5°
Magnetic Declination:	3.3° W
Corrected Azimuth:	061.2°
Difference Calculated vs. Corrected:	0.6°
Measured Inclination:	+13°

Sacred Plaza

Calculated June Solstice Sun Rise Azimuth:	062.2°
Measured Azimuth:	065.0°
Magnetic Declination:	3.3° W
Corrected Azimuth:	061.7°
Difference Calculated vs. Corrected:	0.5°
Measured Inclination:	+12°

Calculated December Solstice Sun Set Azimuth:	246.6°
Measured Azimuth:	248.0°
Magnetic Declination:	3.3° W
Corrected Azimuth:	244.7°
Difference Calculated vs. Corrected:	1.9°
Measured Inclination:	+3°

The Mortars

Calculated Equinox Sun Rise Azimuth:	081.8°
Measured Azimuth:	083.5°
Magnetic Declination:	3.3° W
Corrected Azimuth:	080.2°
Difference Calculated vs. Corrected:	1.6°
Measured Inclination:	+33°

Huayna Picchu

Calculated December Solstice Sun Set Azimuth:	246.1°
Measured Azimuth:	244.5°

Magnetic Declination:	3.5° W
Corrected Azimuth:	241.0°
Difference Calculated vs. Corrected:	5.1°
Measured Inclination:	+1°

Temple of the Moon

Calculated June Solstice Sun Set Azimuth:	298.6°
Measured Azimuth:	295.0°
Magnetic Declination:	3.2° W
Corrected Azimuth:	291.8°
Difference Calculated vs. Corrected:	6.8°
Measured Inclination:	+14°

9.3 River Intihuatana

Tower

Calculated ESR Azimuth:	085.8°
Measured Azimuth:	090.0°
Magnetic Declination:	3.3° W
Corrected Azimuth:	086.7°
Difference Calculated vs. Corrected:	0.9°
Measured Inclination:	+19°

9.4 Llactapata

Sun Temple

Calculated June Solstice Sun Rise Azimuth:	064.5°
Measured Azimuth:	066.5°
Magnetic Declination:	3.3° W
Corrected Azimuth:	063.2°
Difference Calculated vs. Corrected:	1.3°
Measured Inclination:	+5°

Calculated Azimuth to Sacred Plaza:	060.5°
Measured Azimuth:	064.2°
Magnetic Declination:	3.3° W
Corrected Azimuth:	060.9°
Difference Calculated vs. Corrected:	0.4°

A.3 GPS Field Data

Cusco Region, Tipon, and Saihuite

7.1 Kenko Grande

Rear entrance to cave

S13° 30′ 33″ W071° 58′ 14″
3594 masl

Front entrance to cave

S13° 30′ 33″ W071° 58′ 14″
3596 masl

7.2 Kenko Chico

S13° 30′ 37″ W071° 58′ 19″
3590 masl

7.3 Mesa Redonda

S13° 30′ 48″ W071° 58′ 11″
3474 masl

Fig. A.3 Cuy (watercolor by Jessica Gullberg)

7.4 Tetecaca

S13° 30′ 54″ W071° 57′ 56″
3492 masl

7.5 Patallacta

S13° 30′ 45″ W071° 58′ 06″
3543 masl

7.6 Kusilluchayoc

S13° 30′ 29″ W071° 57′ 57″
3625 masl

7.7 Lacco

Northeast Cave

S13° 30′ 20″ W071° 57′ 52″
3655 masl

Southwest Cave

S13° 30′ 21″ W071° 57′ 54″
3662 masl

Southeast Cave

S13° 30′ 22″ W071° 57′ 51″
3648 masl

7.8 Huaca 44

S13° 30′ 18″ W071° 58′ 41″
3662 masl

7.9 Lanlakuyok

S13° 30′ 21″ W071° 58′ 44″
3727 masl

7.10 Puca Pucara

S13° 29′ 0″ W071° 57′ 43″
3805 masl

7.11 Tambomachay

S13° 28′ 35″ W071° 58′ 5″
3900 masl

7.12 Sacsahuaman

Tired Stone

S13° 28′ 35″ W071° 58′ 5″
3610 masl

Throne of the Inca

S13° 30′ 48″ W071° 58′ 11″
3618 masl

Top of zigzag walls

S13° 30′ 26″ W071° 58′ 54″
3607masl

Base of zigzag walls

S13° 30′ 30″ W071° 58′ 55″
3595 masl

7.13 Mollaguanca

S13° 30′ 21″ W071° 58′ 44″
3581 masl

7.14 Sapantiana

S13° 30′ 44″ W071° 58′ 41″
3465 masl

7.15 Rumiwasi Bajo – Tunnel

S13° 31′ 12″ W071° 56′ 26″
3740 masl

7.16 Rumiwasi Alto

S13° 30′ 58″ W071° 56′ 21″
3563 masl

7.17 Kusicallanca

S13° 31′ 13″ W071° 56′ 14″
3573 masl

7.18 Tipon

Iglesia Raqui

S13° 34′ 14″ W071° 46′ 15″
3476 masl

Top of terrace system

S13° 34′ 9″ W071° 46′ 54″

Intihuatana

S13° 24′ 10″ W071° 47′ 8″
3554 masl

7.19 Saihuite

Principle Stone

S13° 32′ 50″ W072° 48′ 11″
3653 masl

Niche toward horizon

S13° 32′ 50″ W072° 48′ 11″
3654 masl

Top of fountain with stairs

S13° 32′ 49″ W072° 48′ 8″
3640 masl

Bottom of fountain with stairs

S13° 32′ 49″ W072° 48′ 7″
3623 masl

Rumihuasi Stone

S13° 32′ 46″ W072° 47′ 58″
3584 masl

Chingana Platform

S13° 32′ 53″ W072° 47′ 54″
3565 masl

Third Stone

S13° 32′ 50″ W072° 47′ 47″
3545 masl

The Sacred Valley

8.1 Chinchero

First rock east of the chapel

S13° 23 24″ W072° 02′ 47″
3790 masl

Kondorkaka

S13° 23′ 21″ W072° 02′ 37″
3773 masl

Mesakaka

S13° 23′ 17″ W072° 02′ 35″
3744 masl

Chinkana

S13° 23′ 16″ W072° 2′ 35″
3724 masl

Titikaka

S13° 23′ 48″ W072° 2′ 48″
3753 masl

8.2 Pisac

S13° 24′ 51″ W071° 50′ 38″
3395 masl

8.3 Q'espiwanka

Boulder

S13° 18′ 5″ W072° 6′ 47″
2922 masl

Triple-jamb doorway

S13° 18′ 4″ W072° 6′ 40″
2933 masl

Outside south wall

S13° 18′ 5″ W072° 6′ 47″
2922 masl

8.4 Cerro Pumahuachana

South wall remains – south end

S13° 19′ 5″ W072° 3′ 54″
3568 masl

South wall remains – north end

S13° 19′ 4″ W072° 3′ 55″
3572 masl

North wall remains – south end

S13° 19′ 2″ W072° 3′ 55″
3575 masl

North wall remains – north end

S13° 18′ 59″ W072° 3′ 56″
3592 masl

Below chapel

S13° 18′ 50″ W072° 3′ 56″
3618 masl

8.5 Cerro Unoraqui

Village on side of mountain

S13° 20′ 6″ W071° 59′ 20″
2937 masl

Mountaintop

S13° 20′ 21″ W072° 0′ 27″
4377 masl

8.6 Maras

S13° 17′ 55″ W072° 9′ 24″
3380 masl

8.7 Moray

S13° 19′ 49″ W072° 11′ 45″
3385 masl

8.8 Choquequilla

S13° 17′ 32″ W072° 13′ 56″
3627 masl

8.9 Ollantaytambo

S13° 15′ 17″ W072° 15′ 48″
2792 masl

Machu Picchu Region

9.2 Machu Picchu

Intihuatana

S13° 9′ 47″ W072° 32′ 44″
2468 masl

Sacred Plaza

S13° 9′ 49″ W072° 32′ 44″
2466 masl

Inca Bridge

S13° 10′ 13″ W072° 32′ 40″
2550 masl

Mortars

S13° 9′ 48″ W072° 32′ 40″
2460 masl

Huayna Picchu – mountaintop

S13° 9′ 24″ W072° 32′ 45″
2698 masl

Huayna Picchu – Building of Three Windows

S13° 9′ 24″ W072° 32′ 45″
2690 masl

Temple of the Moon

S13° 9′ 5″ W072° 32′ 47″
2277 masl

9.3 River Intihuatana

S13° 10′ 32″ W072° 33′ 26″
1819 masl

9.4 Llactapata

Base Camp

S13° 11 10″ W072° 34 53″
2601 masl

Sun Temple

S13° 11′ 9″ W072° 35′ 5″
2740 masl

A.4 Magnetic Declination

Cusco Region, Tipon, and Saihuite

7.1 Kenko Grande

S13° 30′ 31″ W071° 58′ 14″
10/26/063° 35′ W

7.2 Kenko Chico

S13° 30′ 37″ W071° 58′ 19″
6/5/08 3° 51′ W

7.3 Mesa Redonda

S13° 30′ 48″ W071° 58′ 11″
10/17/083° 55′ W

7.4 Tetecaca

S13° 30′ 54″ W071° 57′ 56″
10/17/083° 55′ W

Fig. A.4 Traditional corn (watercolor by Jessica Gullberg)

7.5 Patallacta

S13° 30′ 45″ W071° 58′ 06″
10/17/083° 55′ W

7.6 Kusilluchayoc

S13° 30′ 29″ W071° 57′ 57″
10/17/083° 55′ W

7.7 Lacco

S13° 30′ 21″ W071° 57′ 51″
10/17/083° 55′ W

7.8 Huaca 44

S13° 30′ 18″ W071° 58′ 41″
10/17/083° 55′ W

7.9 Lanlakuyok

S13° 30′ 21″ W071° 58′ 44″
10/14/083° 55′ W

7.10 Puca Pucara

S13° 29′ 0″ W071° 57′ 43″
10/14/083° 55′ W

7.11 Tambomachay

S13° 28′ 35″ W071° 58′ 5″
10/17/083° 55′ W

7.12 Sacsahuaman

Tired Stone

S13° 30′ 15″ W071° 58′ 52″
10/14/083° 54′ W

Throne of the Inca

S13° 30′ 48″ W071° 58′ 11″
10/14/083° 54′ W

7.13 Mollaguanca

S13° 30′ 21″ W071° 58′ 44″
10/14/083° 54′ W

7.14 Sapantiana

S13° 30′ 44″ W071° 58′ 41″
10/14/083° 54′ W

7.15 Rumiwasi Bajo

S13° 31′ 12″ W071° 56′ 26″
10/14/083° 55′ W

7.16 Rumiwasi Alto

S13° 30′ 58″ W071° 56′ 21″
10/16/083° 56′ W

7.17 Kusicallanca

S13° 31′ 13″ W071° 56′ 14″
10/16/083° 56′ W

7.18 Tipon

S13° 34′ 14″ W071° 46′ 15″
10/16/084° 3′ W

7.19.2 Saihuite – Upper Structure

S13° 32′ 50″ W072° 48′ 11″
10/15/083° 13′ W

The Sacred Valley

8.1 Chinchero

Chinkana

S13° 23′ 16″ W072° 2′ 35″
10/25/073° 43′ W

Titikaka

S13° 23′ 48″ W072° 2′ 48″
10/25/073° 43′ W

8.2 Pisac

S13° 24′ 51″ W071° 50′ 38″
10/16/084° 2′ W

8.3 Q’espiwanka

S13° 18′ 5″ W072° 6′ 47″
6/8/08 3° 47′ W

8.4 Cerro Pumahuachana

S13° 19′ 4″ W072° 3′ 55″
6/7/08 3° 49′ W

8.5 Cerro Unoraqui

S13° 20′ 21″ W072° 0′ 27″
10/17/083° 55′ W

8.6 Maras

S13° 17′ 55″ W072° 9′ 24″
9/12/18 5° 36′ W

8.7 Moray

S13° 19′ 49″ W072° 11′ 45″
9/12/18 3° 34′ W

8.8 Choquequilla

S13° 17′ 32″ W072° 13′ 56″
10/17/083° 55′ W

8.9 Ollantaytambo

S13° 15′ 17″ W072° 15′ 48″
10/27/06 3° 25′ W

Machu Picchu Region

9.2 Machu Picchu

Intihuatana

S13° 9′ 47″ W072° 32′ 44″
6/19/07 3° 19′ W

Sacred Plaza

S13° 9′ 49″ W072° 32′ 44″
6/19/07 3° 19′ W

The Mortars

S13° 9′ 48″ W072° 32′ 40″
6/19/07 3° 19′ W

Huayna Picchu

S13° 9′ 24″ W072° 32′ 45″
6/6/08 3° 29′ W

Temple of the Moon

S13° 9′ 5″ W072° 32′ 47″
10/28/063° 13′ W

9.3 River Intihuatana

S13° 10′ 32″ W072° 33′ 26″
6/17/07 3° 18′ W

9.4 Llactapata – Sun Temple

S13° 11′ 9″ W072° 35′ 5″
6/18/07 3° 17′ W

A.5 Solar Horizon Positions

Angular positions calculated from Appendix A.6 formula (A.8)

Cusco Region, Tipon, and Saihuite

7.1 Kenko Grande

S13° 30′ 31″ W071° 58′ 14″
ESR 2° up; ESS 6° up

June 21

ESR	089.93°
ESS	271.03°

7.2 Kenko Chico

S13° 30′ 37″ W071° 58′ 19″
DSSR 2° up

December 21

DSSR	113.63°

Fig. A.5 Peru Rail to Aguas Calientas and Machu Picchu (watercolor by Jessica Gullberg)

7.6 Kusilluchayoc

S13° 30′ 29″ W071° 57′ 57″
JSSS 4° up

June 21

JSSS	295.26°

7.7 Lacco

S13° 30′ 21″ W071° 57′ 51″
JSSR 4° up

June 22

JSSR	064.74°

7.8 Huaca 44

S13° 30′ 18″ W071° 58′ 41″
JSSR 3° up
ESR 2° up
DSSR 2° up

June 21

JSSR	065.03°

March 20

ESR	089.56°

December 21

DSSR	113.63°

7.9 Lanlakuyok

S13° 30′ 21″ W071° 58′ 44″
ESR 2° up

March 20

ESR	089.56°

7.10 Puca Pucara

S13° 29′ 0″ W071° 57′ 43″
ESR 3° up
ESS 4° up

March 20

ESR	089.69°
ESS	270.55°

7.11 Tambomachay

S13° 28′ 35″ W071° 58′ 5″
DSSR 1° up

December 21

DSSR	113.88°

7.13 Mollaguanca

S13° 30′ 21″ W071° 58′ 44″
JSSR 3° up
DSSS 2° up

June 21

JSSR	065.03°

December 21

DSSS	246.37°

7.15 Rumiwasi Bajo

S13° 31′ 12″ W071° 56′ 26″
DSSR 2° up

December 21

DSSR	113.63°

7.16 Rumiwasi Alto

S13° 30′ 58″ W071° 56′ 21″
DSSR 1° up

December 21

DSSR	113.88°

7.17 Kusicallanca

S13° 31′ 13″ W071° 56′ 14″
DSSR 0°

December 21

DSSR	114.14°

7.18 Tipon

S13° 34′ 14″ W071° 46′ 15″
JSSS Iglesia Raqui 13° up
JSSS Intihuatana 1° up

June 21

JSSS Iglesia Raqui	298.37°
JSSS Intihuatana	294.42°

7.19 Saihuite – Upper Structure Niche

S13° 32′ 50″ W072° 48′ 11″
JSSR 2° up

June 21

JSSR	065.31°

The Sacred Valley

8.1 Chinchero

Chinkana

S13° 23′ 16″ W072° 2′ 35″

DSSS 10° up

December 21

DSSS	248.09°

Titikaka

S13° 23′ 48″ W072° 2′ 48″
JSSR 5° up

June 21

JSSR	064.46°

8.2 Pisac

S13° 24′ 51″ W071° 50′ 38″
JSSR 0°
DSSR 0°

June 21

JSSR	065.87°

December 21

DSSR	114.13°

8.3 Q'espiwanka

S13° 18′ 5″ W072° 6′ 47″
JSSR Outside Palace 21° up
JSSR Boulder 22° up
JSSS 5° up
DSSR 6° up
DSSS 5° up
ESR 14° up
ESS 5° up
ZSR 4° up
AZSS 5° up

June 21

JSSR Outside Palace	58.87°
JSSR Boulder	57.57°
JSSS	295.52°

March 20

ESR	87.05°
ESS	270.77°

December 21

DSSR	112.69°
DSSS	247.31°

February 13

ZSR	103.08°

August 18

AZSS	284.95°

8.8 Choquequilla

S13° 17′ 32″ W072° 13′ 56″
DSSR 32° up

December 21

DSSR	109.53°

Machu Picchu Region

9.2 Machu Picchu

Intihuatana

S13° 9′ 47″ W072° 32′ 44″

JSSR 13° up

June 21

JSSR	061.76°

Sacred Plaza

S13° 9′ 49″ W072° 32′ 44″
JSSR 12° up

June 21

JSSR	062.16°

The Mortars

S13° 9′ 48″ W072° 32′ 40″
ESR 33° up

March 20

ESR	081.76°

Huayna Picchu

S13° 9′ 24″ W072° 32′ 45″
DSSS 1° up

December 21

DSSS	246.15°

Temple of the Moon

S13° 9′ 5″ W072° 32′ 47″
JSSS 14° up

June 21

JSSS	298.63°

9.3 River Intihuatana

S13° 10′ 32″ W072° 33′ 26″

ESR 19° up

March 20

ESR	085.81°

9.4 Llactapata – Sun Temple

S13° 11′ 9″ W072° 35′ 5″
JSSR 5° up

June 21

JSSR	064.50°

A.6 Spherical Trigonometry Formulae

If a spherical triangle is defined as having three angles labeled A, B and C, and the sides opposite those angles are correspondingly labeled a, b and c, then as shown in Smart (1977) the following are the basic formulae for solving such a triangle:

$$\cos a = \cos b \cos c + \sin b \sin c \cos A \tag{A.1}$$

$$\sin A / \sin a = \sin B / \sin b = \sin C / \sin c \tag{A.2}$$

$$\sin a \cos B = \cos b \sin c - \sin b \cos c \cos A \tag{A.3}$$

$$\cos a \cos C = \sin a \cot b - \sin C \cot B \tag{A.4}$$

Aveni (1980) gives specific formulae useful for calculating specific horizon points of celestial bodies:

$$\textit{Hour angle of the Sun in degrees} = (GMT\text{-}12)15 - LONG - (EOT)15 \tag{A.5}$$

$$\textit{Altitude of the Sun} = Arcsin(\sin(LAT)\sin(DEC) + \cos(LAT)\cos(DEC)\cos HA) \tag{A.6}$$

$$\textit{Azimuth of the Sun} = Arcsin(\sin(HA)\cos(DEC)/\cos(ALT)) \tag{A.7}$$

The following formula is most useful for calculating the actual position on the horizon that the Sun will rise due to horizon elevation:

Fig. A.6 Elevated horizon viewed from the Llactapata Sun Temple (pen and watercolor by Jessica Gullberg)

$$\text{Azimuth of Sun} = \text{Arccos}\left(\sin(DEC) - \sin(LAT)\sin(ALT)\right)/\left(\cos(LAT)\cos(ALT)\right) \tag{A.8}$$

HA = Hour Angle
GMT = Universal Time Coordinated
LONG = Longitude
EOT = Equation of Time
LAT = Latitude
DEC = Declination
ALT = Altitude

Bibliography

Adkins, L. R. and Benfer, R. A., 2009. Lunar Standstill Markers at Preceramic Temples at the Buena Vista Site in Perú. In Rubiño-Martín, J. A., Belmonte, J. A., Prada, F., and Alberdi, A. (eds.), *Cosmology Across Cultures,* ASP Conference Series, 409. San Francisco, Astronomical Society of the Pacific, 267–271.

Arévalo, J., 2002. *Inka Initiation Path: The Awakening of the Puma, Evidences of Archaeo-astronomy in the Andes.* Cusco, Shamanic Productions.

Arévalo, J., 2007. *Inka Power Places: Solar Initiation, Andean Archaeo-astronomy.* Cusco, Shamanic Productions.

Arriaga, P. J., 1968 [1621]. *The Extirpation of Idolatry in Peru.* Keating, L. C. (ed. and trans.). Lexington, University of Kentucky Press.

Aveni, A. (ed.), 1975, *Archaeoastronomy in Pre-Columbian America.* Austin, University of Texas Press.

Aveni, A. (ed.), 1977. *Native American Astronomy.* Austin, University of Texas Press.

Aveni, A., 1980. *Skywatchers of Ancient Mexico.* Austin, University of Texas Press.

Aveni, A., 1981a. Horizon Astronomy in Incaic Cusco. In R. Williamson (ed.) *Archaeoastronomy in the Americas.* Los Altos and College Park, Ballena Press, 305–318.

Aveni, A., 1981b. Archaeoastronomy. *Advances in Archaeological Method and Theory,* 4, 1–77.

Aveni, A., 1981c. Tropical Archaeoastronomy. *Science,* 213, 4504, 161–171.

Aveni, A., 1981d. Comment on J. H. Rowe: Archaeoastronomy in Mesoamerica and Peru. *Latin American Research Review,* 16, 3, 163–166.

Aveni, A., 1987. On Seeing the Light. Reply to David Dearborn and Katharina Schreiber: Here Comes the Sun: The Cuzco-Machu Picchu Connection. *Archaeoastronomy,* X, 22–24.

Aveni, A., 1989. *Empires of Time.* New York, Basic Books.

Aveni, A., 1996. Astronomy in the Americas. In Walker, C. (ed.) *Astronomy Before the Telescope.* New York, St. Martin's Press.

Aveni, A., 1997. *Stairways to the Stars: Skywatching in Three Great Ancient Cultures.* New York, John Wiley & Sons.

Aveni, A., 1999. Review of Brian S. Bauer: *The Sacred Landscape of the Inca: The Cusco Ceque System. Latin American Antiquity,* 10, 3, 323–324.

Aveni, A., 2002. *Empires of Time.* Boulder, University Press of Colorado.

Aveni, A. (ed.), 2008. *Foundations of New World Cultural Astronomy: A Reader with Commentary.* Boulder, University Press of Colorado.

Aveni, A. (ed.), 2009. *Archaeoastronomy in the New World: American Primitive Astronomy.* Cambridge, Cambridge University Press.

Aveni, A. & Urton G. (eds.), 1982. *Ethnoastronomy and Archaeoastronomy in the American Tropics.* New York, New York Academy of Sciences.

© Springer Nature Switzerland AG 2020

S. R. Gullberg, *Astronomy of the Inca Empire*, Historical & Cultural Astronomy,
https://doi.org/10.1007/978-3-030-48366-1

Bauer, B., 1992. Ritual Pathways of the Inca: An Analysis of the Collasuyu Ceques in Cuzco. *Latin American Antiquity,* 3, 3, 183–205.
Bauer, B., 1996. Legitimization of the State in Inca Myth and Ritual. *American Anthropologist,* 98, 2, 327–337.
Bauer, B., 1998. *The Sacred Landscape of the Inca: The Cusco Ceque System.* Austin, University of Texas Press.
Bauer, B., 2004. *Ancient Cuzco: Heartland of the Inca.* Austin, University of Texas Press.
Bauer, B. and Dearborn, D., 1995. *Astronomy and Empire in the Ancient Andes: the Cultural Origins of Inca Sky Watching,* Austin, University of Texas Press.
Bauer, B. and Stanish, C., 2001. *Ritual and Pilgrimage in the Ancient Andes.* Austin, University of Texas Press.
Bely, P., Ed., 2003. *The Design and Construction of Large Optical Telescopes.* New York, Springer.
Benson, E. P. and Cook, A. G., 2001. *Ritual Sacrifice in Ancient Peru.* Austin, University of Texas Press.
Betanzos, J., 1996 [1576]. *Narrative of the Incas.* Buchanan, D. (ed.) and Hamilton, R. (trans.). Austin, University of Texas Press.
Bingham, H. 1930. *Machu Picchu: A Citadel of the Incas.* New Haven, Yale University Press.
Bingham, H., 1948. *Lost City of the Incas: The Story of Machu Picchu and its Builders.* New York, Phoenix.
Bray, T. L., 2009. An Archaeological Perspective on the Andean Concept of *Camaquen*: Thinking Through Late Pre-Columbian *Ofrendas* and *Huacas. Cambridge Archaeological Journal,* 19, 3, 357–366.
Bray, T. L., 2015. Andean Wak'as and Alternative Configurations of Persons, Power, and Things. In T. L. Bray, *The Archaeology of Wak'as: Explorations of the Sacred in the Pre-Columbian Andes* (pp. 3–19). Boulder: University Press of Colorado.
Burger, R. L., 1989. An Overview of Peruvian Archaeology (1976-1986). *Annual Review of Anthropology,* 18, 37–69.
Burger, R. L., 1992. *Chavin and the Origins of Andean Civilization.* London, Thames and Hudson.
Burger, R. L. and Matos M. R.., 2002. Atalla: A Center on the Periphery of the Chavin Horizon. *Latin American Antiquity,* 13, 2, 153–177.
Burger, R. L. and Salazar-Burger, L., 1991. The Second Season of Investigations at the Initial Period of Cardal, Peru, *Journal of Field Archaeology,* 18, 3, 275–296.
Burger, R. L. and Van Der Merwe, N. J., 1990. Maize and the Origin of Highland Chavín Civilization: An Isotopic Perspective. *American Anthropologist,* 92, 1, 85–95.
Chauvenet, W., 1960. *A Manual of Spherical and Practical Astronomy,* Vols 1 and 2. New York, Dover.
Christie, J. J., 2007. Did the Inka Copy Cusco? An Answer Derived from an Architectural-Sculptural Model. *Journal of Latin American and Caribbean Anthropology, 12*(1), 164–199.
Christie, J. J., 2013. Inka Iconographic Rocks. *IFRAO 2013 Proceedings, American Indian Rock Art, 40.*
Cieza de León, P., 1998 [1555]. *The Discovery and Conquest of Peru.* Parma-Cook, A. and Cook, N. (eds. and trans.). Chapel Hill, Duke University Press.
Cobo, B., 1983 [1653]. *History of the Inca Empire: An Account of the Indians' Customs and Their Origin, Together with a Treatise on Inca Legends, History, and Social Institutions.* Hamilton, R. (ed. and trans.). Austin, University of Texas Press.
Cobo, B., 1990 [1653]. *Inca Religion and Customs.* Hamilton, R. (ed. and trans.). Austin, University of Texas Press.
Conklin, W. J.., 1982. The Information System of Middle Horizon Quipus. In Aveni, A. & Urton G. (eds.). *Ethnoastronomy and Archaeoastronomy in the American Tropics.* New York, New York Academy of Sciences, 261–281.
Cummins, T. B. F., 1992. Review of R. Tom Zuidema: *Inca Civilization in Cuzco.* In *American Anthropologist,* 94, 1, 222.
D'Altroy, T., 2002. *The Incas.* Hoboken, NJ, Wiley-Blackwell.

Dearborn, D., 1987a. Blinded by the Light. Response to Anthony Aveni: On Seeing the Light. *Archaeoastronomy,* X, 24–28.

Dearborn, D., 1987b. Stand and Deliver. Response to R. Tom Zuidema's reply to David Dearborn and Katharina Schreiber: Here Comes the Sun: The Cuzco-Machu Picchu Connection. *Archaeoastronomy,* X, 24–28.

Dearborn, D. and Bell, B., 1984. New Tools for Ancient Skies. *Archaeoastronomy,* VII, 1–4, 96–104.

Dearborn, D. and Schreiber, K., 1986. Here Comes the Sun: The Cusco-Machu Picchu Connection. *Archaeoastronomy,* IX, 15–36.

Dearborn, D., Schreiber, K. and White, R., 1987. Intimachay, A December Solstice Observatory. *American Antiquity,* 52, 346–352.

Dearborn, D., Seddon, M., and Bauer, B., 1998. The Sanctuary of Titicaca: Where the sun returns to earth. *Latin American Antiquity,* 9, 3, 240–258.

Dearborn, D. and White, R., 1982. Archaeoastronomy at Machu Picchu. In Aveni, A. & Urton G. (eds.). *Ethnoastronomy and Archaeoastronomy in the American Tropics.* New York, New York Academy of Sciences, 249–259.

Dearborn, D. and White, R., 1983. The "Torreon" at Machu Picchu as an Observatory. *Archaeoastronomy* 5, Supplement of the *Journal for History of Astronomy*, S37-S49.

Dearborn, D, and White, R., 1989. Inca Observatories: Their Relation to the Calendar and Ritual. In Aveni, A. (ed.). *World Archaeoastronomy*. Cambridge, Cambridge University Press, 462–469.

DeLeonardis L. and Lau, G., 2004. Life, Death, and Ancestors. Silverman, H. (ed.). *Andean Archaeology.* Malden, MA, 77–115.

Eliade, M., 1972. *Shamanism: Archaic Techniques of Ecstasy*. Princeton, Princeton University Press.

Evans, J., 1998. *The History and Practice of Ancient Astronomy*. Oxford, Oxford University Press.

Farrington, I. S., 1992. Ritual Geography, Settlement Patterns and the Characterization of the Provinces of the Inka Heartland. *World Archaeology,* 23, 3, 368–385.

Farrington, I. S., 1995. The Mummy, Palace, and Estate of Inka Huayna Capac at Quispeguanca. *Tawantinsuyu,* I, 55–65.

Farrington, I. S. 2017. Aspects of the sacred and kingship at the Inka palace of Qespiwanka. *Time and Mind, 10*(2), 1–40.

Frost, P., 1989. *Exploring Cuzco*, 4th Cusco, Nuevas Imaìgenes.

Garcilaso de la Vega, 1961 [1609]. *The Royal Commentaries of the Inca*. New York, Avon.

Gamara, M. R., 2017. *Matemática Inka*. Cusco, Líder Digital.

Gasparini, G. and Margolies, L., 1980. *Inca Architecture*. Bloomington, Indiana University Press.

Ghezzi, I., 2006. Religious Warfare at Chankillo. In Isbell, W. H. and Silverman, H. (eds.), *Andean Archaeology III: North and South*. New York, Springer, 67–84.

Guaman Poma de Ayala, F., 2009 [1615]. *The First New Chronicle and Good Government: On the History of the World and the Incas up to 1615*. Hamilton, R. (ed. and trans.). Austin, University of Texas Press.

Gullberg, S., 2010. Inca Solar Orientations in Southeastern Peru. *Journal of Cosmology*, 9, 2078–2091.

Gullberg, S., 2015. Marking Time in the Inca Empire. *Journal of Skyscape Archaeology,* 1, 2, 217–241.

Gullberg, S., 2016. Astronomy and the Ceques of Cusco. In F. Silva, J. M. Malville, T. Lomsdalen, & F. Ventura (eds.), *The materiality of the sky*. Lampeter, United Kingdom: Sophia Center Press, 255–266.

Gullberg, S., 2018. Inca Astronomy: Horizon, Light, and Shadow. *Astronomische Nachrichten*. 340, 1–3, 23–29.

Gullberg, S. and Malville, J. M., 2011. The Astronomy of Peruvian Huacas. In Orchiston, W., Nakamura, T., and Strom, R. (eds.), *Highlighting the History of Astronomy in the Asia-Pacific Region*. New York, Springer.

Gullberg, S. and Malville, J. M., 2014. The River Intihuatana: Huaca Sanctuary on the Urubamba. *Mediterranean Archaeology &Archaeometry*, 14, 3, 179–187.

Gullberg, S., and Malville, J. M., 2017. Caves, Liminality, and the Sun in the Inca World. *Culture and Cosmos*, 21, 1 & 2, 193–214.

Haas, J. and Creamer, W., 2004. Cultural Transformations in the Central Andean Late Archaic. In Silverman, H. (ed.). *Andean Archaeology.* Malden, MA, Blackwell, 35–50.

Hadingham, E., 1984. *Early Man and the Cosmos.* Norman, University of Oklahoma Press.

Hanzalova, K., Klokocnik, J., and Kostelecký, J., 2014. Astronomical orientations of Incas objects in Machu Picchu and Ollantaytambo. *ISPRS – International Archives of the Photogrammetry, Remote Sensing and Spatial Information Sciences, XL-5*, 273–276.

Hanzalova, K., Klokocnik, J. and Kostelecký, J., 2015. New discoveries on astronomical orientation of Inca site in Ollantaytambo, Peru. *Geoinformatics FCE CTU, 14*(2), 37–46.

Hawkins, G. S., 1975. Astroarchaeology: The Unwritten Evidence. In Aveni, A. F. (ed.), *Archaeoastronomy in Pre-Columbian America*. Austin, University of Texas Press.

Hemming, J., 1970. *The Conquest of the Incas*. San Diego, Harcourt.

Hemming, J. and Ranney, E., 1982. *Monuments of the Incas.* Albuquerque, University of New Mexico Press.

Isbell, W. H. and Vranich, A., 2004. Experiencing the Cities of Wari and Tiwanaku. Silverman, H. (ed.). *Andean Archaeology.* Malden, MA, Blackwell, 167–182.

Kelley, D. and Milone, E., 2005. *Exploring Ancient Skies: An Encyclopedic Survey of Archaeoastronomy.* New York, Springer.

Kembel, S. R. and Rick, J. W., 2004. Building Authority at Chavin de Huántar: Models of Social Organization and Development in the Initial Period and Early Horizon. Silverman, H. (ed.). *Andean Archaeology.* Malden, MA, Blackwell, 35–76.

Kment, T., Rauter, M., and Zotti, G., 2006. Modelling of Daylight for Computer Graphics. Forschungsseminar ICGA (SS2006). Retrieved from https://www.academia.edu/809995/Modelling_of_Daylight_for_Computer_Graphics

Krupp, E. C., 1983. *Echoes of the Ancient Skies: The Astronomy of Lost Civilizations.* New York, Harper & Row.

Krupp, E. C., 1991. *Beyond the Blue Horizon: Myths and Legends of the Sun, Moon, Stars and Planets*. New York, Oxford University Press.

Krupp, E. C., 2008. Weathering the Stars. *Sky & Telescope, 115*(6), 43–45.

Lee, V., 2000. *Forgotten Vilcabamba.* Cortez, CO, Sixpac Manco.

Magli, G., 2005. On the Astronomical Content of the Sacred Landscape of Cusco in Inka Times. *Nexus Network Journal, 7*(2), 22–32.

Malville, J. M., 2009. Animating the Inanimate: Camay and Astronomical Huacas of Peru. In Rubiño-Martín, J. A., Belmonte, J. A., Prada, f., and Alberdi, A. (eds.), *Cosmology Across Cultures,* ASP Conference Series, 409. San Francisco, Astronomical Society of the Pacific, 261–266.

Malville, J. M. 2015. Astronomy of Inca Royal Estates I: The Sacred Valley. In C. L. N. Ruggles (Ed.), *Handbook of Archaeoastronomy and Ethnoastronomy* (pp. 869–872). New York, NY: Springer.

Malville, J. M., Thomson, H., and Ziegler, G., 2006. The Sun Temple of Llactapata and the Ceremonial Neighborhood of Machu Picchu. In Bostwick T. & Bates, B. (eds.), *Viewing the Sky Through Past and Present Cultures.* Phoenix, City of Phoenix Parks, Recreation and Library, 327–339.

Malville, J. M., Zawaski, M. and Gullberg, S., 2008. Cosmological Motifs of Peruvian Huacas. In Vaiškūnas, J. (ed.), *Astronomy and Cosmology in Folk Traditions and Cultural Heritage, Archaeologia Baltica, 10.* Vilnius, Klaipėda University Press, 175–182.

Meddens, F. M., Branch, N. P., Pomacanchari, C. V., Riddiford, N., and Kemp, R., 2008. High Altitude *Ushnu* Platforms in the Department of Ayacucho Peru, Structure, Ancestors and Animating Essence.... In Staller, J. E. (ed.), *Pre-Columbian Landscapes of Creation and Origin*. New York, Springer, 315–356.

Medina Zevallos, O., 1998. *The Enigma of MachuPicchu.* Cusco, Milenium-Editores.

Meeus, J., 1991. *Astronomical Algorithms.* Richmond, VA, Willmann-Bell.

Meeus, J., 1997. *Mathematical Astronomy Morsels.* Richmond, VA, Willmann-Bell.

Miller, G. R. and Burger, R. L., 1995. Our Father the Cayman, Our Dinner the Llama: animal Utilization at Chavin de Huantur, Peru. *American Antiquity,* 60, 3, 421–458.

Morris, C. and Thompson, D. E., 1985. *Huanaco Pampa: An Ancient City and its Hinterland.* London, Thames and Hudson.

Niles, S., 1987. *Callachaca: Style and Status in an Inca Community.* Iowa City, University of Iowa Press.

Niles, S., 1991. Review of John Hyslop: *Inka Settlement Planning* and review of R. Tom Zuidema: *Inca Civilization in Cuzco.* In *Journal of Field Archaeology,* 18, 4, 507–511.

Niles, S., 1999. *The Shape of Inca History: Narrative and Architecture in an Andean Empire.* Iowa City, University of Iowa Press.

Niles, S., 2004. The Nature of Inca Royal Estates. In Burger, L. & Salazar, L. (eds.), *Machu Picchu, Unveiling the Mystery of the Incas.* New Haven, Yale University Press, 49–70.

Orlove, B. S., Chiang, J. C., and Cane, M. A., 2000. Forecasting Andean rainfall and crop yield from the influence of El Niño on Pleiades visibility. *Nature,* 403, 68–71.

Paternosto, C., 1989. *The Stone and the Thread: Andean Roots of Abstract Art.* Austin, University of Texas Press.

Pizzaro, P., 1921 [1571]. *Relation of the Discovery and Conquest of the Kingdoms of Peru.* Charleston, SC, BiblioLife.

Polo de Ondegardo, J., 1965 [1571]. A report on the basic principles explaining the serious harm which follows when the traditional rights of the Indians are not respected. Translated by Brunel, A., Murra, J., and Muirden, S. New Haven, Human Relations Area Files, 67.

Reinhard, J., 1985. Sacred Mountains: An Ethno-Archaeological Study of High Andean Ruins. *Mountain Research and Development,* 5, 4, 299–317.

Reinhard, J., 2002 *Machu Picchu, The Sacred Center.* Cusco, Instituto Machu Picchu.

Reinhard, J., 2007. *Machu Picchu: Exploring an Ancient Sacred Center,* 4th ed. Los Angeles, Cotsen Institute of Archaeology.

Reinhard, J. and Ceruti, C., 2005. Sacred Mountains, Ceremonial Sites, and Human Sacrifice Among the Incas. *Archaeoastronomy,* XIX, 1–43.

Roe, P. G., 1996. Mythic Substitution and the Stars: Aspects of Shipibo and Quechua Ethnoastronomy Compared. In Del Chamberlain, V., Carlson, J. B., and Young, M. J. (eds.) *Songs from the Sky.* Bognor Regis, W. Sussex, Ocarina Books, 193–228.

Rowe, J., 1946. *Inca Culture at the Time of the Spanish Conquest.* Washington, USGPO.

Rowe, J., 1979. Review of Anthony Aveni (ed.): *Archaeoastronomy in Pre-Columbian America. Latin American Research Review,* 14, 2, 227–233.

Rowe, J., 1980. Inca Policies and Institutions Relating to the Cultural Unification of the Empire. In Collier, G., Rosaldo, R., and Wirth, J. (eds.), *The Inca and Aztec States, 1400–1800.* New York, Academic Press, 93–118.

Rowe, J., 1981. Reply: Archaeoastronomy in Mesoamerica and Peru. *Latin American Research Review,* 16, 3, 171–172.

Rowe, J., 1990. Machu Picchu a la luz de documentos de siglo XVI. *Histórica,* 14, I, 139–154.

Salazar, F. E. E., and Salazar, E. E., 2014. *Cusco and the Sacred Valley of the Incas* (2nd). Cusco, Peru, Tankar E.I.R.L.

Salazar, L., 2004. Machu Picchu: Mysterious Royal Estate in the Cloud Forest. Burger, L. and Salazar, L. (eds.), *Machu Picchu, Unveiling the Mystery of the Incas.* New Haven, Yale University Press, 21–48.

Salomon, F., 1982. Andean Ethnology in the 1970s: A Retrospective. *Latin American Research Review,* 17, 2, 75–128.

Salomon, F., 1985. The Historical Development of Andean Ethnology. *Mountain Research and Development,* 5, 1, 79–98.

Salomon, F. and Urioste, G., 1991. Introductory Essay in *The Huarochiri Manuscript: A Testament of Ancient and Colonial Andean Religion*. Austin, University of Texas Press.
Sarmiento de Gamboa, P., 2009 [1572]. *History of the Incas*. Bauer, B. and Smith, V. (trans.). Charleston, SC, BiblioBazaar.
Schaefer, B. E., 1986. Atmospheric Extinction Effects on Stellar Alignments. *Archaeoastronomy* 10, Supplement of the *Journal for History of Astronomy*, S32.
Schaefer, B. E., 1993. Astronomy and the Limits of Vision. *Archaeoastronomy*, XI, 78–90.
Schaefer, B. E., 2000. New Methods and Techniques for Historical Astronomy and Archaeoastronomy. *Archaeoastronomy*, XV, 121–136.
Sherbondy, J., 1992. Water Ideology in Inca Ethnogenesis. In Dover, R. V. H., Seibold, K. E., and McDowell, J. H. (eds.) *Andean Cosmologies through Time: Persistence and Emergence*. Bloomington, Indiana University Press, 46–66.
Silverman, H., 1994. The Archaeological Identification of an Ancient Peruvian Pilgrimage Center. *World Archaeology*, 26, 1, 1–18.
Silverman, H., 2000. Review of Susan A. Niles: *The Shape of Inca History. Narrative and Architecture in an Andean Empire. Journal of Field Archaeology*, 27, 2, 230–233.
Silverman, H., 2004. Introduction: Space and Time in the Central Andes. Silverman, H. (ed.). *Andean Archaeology*. Malden, MA, Blackwell, 1–15.
Smart, W., 1977. *Spherical Astronomy*. Cambridge, Cambridge University Press.
Solis, R. S., 2006. America's First City? The Case of Late Archaic Caral. Isbell, W. H. and Silverman, H. (eds.). *Andean Archaeology III: North and South*. New York, Springer, 28–66.
Squire, E. G., 1878. *Peru: Incidents of Travel and Exploration in the Land of the Incas*. London, Macmillan and Co.
Staller, J. E., 2008. Dimensions of Place: The Significance of Centers to the Development of Andean Civilization: An Exploration of the *Ushnu* Concept. In Staller, J. E. (ed.), *Pre-Columbian Landscapes of Creation and Origin*. New York, Springer, 269–314.
Stanish, C., 1994. The Hydraulic Hypothesis Revisited: Lake Titicaca Basin Raised Fields in Theoretical Perspective. *Latin American Antiquity*, 5, 4, 312–332.
Stanish, C., 2001. The Origin of State Societies in South America. *Annual Review of Anthropology*, 30, 41–64.
Stastny, F., 1989. El Arte de la Nobleza Inca y La Identidad Andina. In Lavin, I. (ed.) *WORLD ART: Themes of Unity in Diversity. Acts of the XXVIth International Congress of the History of Art*. University Park, PA, Penn State Press, 731–738.
Sullivan, W., 1996. *The Secret of the Incas: Myth, Astronomy, and the War Against Time*. New York, Three Rivers Press.
Taylor, G., 1974. Camay, camac, et camasca dans le manuscript quechua de Huarochiri. *Journal de la Societe des americanistes*, 63, 231–243.
Tello, J. C., 1943. Discovery of the Chavin Culture in Peru. *American Antiquity*, 9, 1, 135–160.
Thomson, H., 2001. *The White Rock: An Exploration of the Inca Heartland*. London, Weidenfeld and Nicolson.
Thomson, H., 2006. *A Sacred Landscape: The Search for Ancient Peru*. Woodstock, Overlook Press.
Titu Cusi Yupanqui, 2005 [1570]. *An Inca Account of the Conquest of Peru*. Bauer, R. (trans.). Boulder, University Press of Colorado.
Titu Cusi Yupanqui, 2006 [1570]. *History of How the Spaniards Arrived in Peru*. Julien, C. (trans.). Indianapolis, Hackett.
Urton, G., 1981a. *At the Crossroads of Earth and Sky: An Andean Cosmology*. Austin, University of Texas Press.
Urton, G., 1981b. The Use of Native Cosmologies in Arcaeoastronomical Studies: The View from South America. In Williamson, R. (ed.) *Archaeoastronomy in the Americas*. Los Altos and College Park: Ballena Press, 285–304.

Urton, G., 1981c. Animals and Astronomy in the Quechua Universe. *Proceedings of the American Philosophical Society*, 125, 2, 110–127.
Urton, G., 1982. Astronomy and Calendrics on the Coast of Peru. In Aveni, A. & Urton G. (eds.). *Ethnoastronomy and Archaeoastronomy in the American Tropics*. New York, New York Academy of Sciences, 231–247.
Urton, G., 1992. Communalism and Differentiation in an Andean Community. In Dover, R. V. H., Seibold, K. E., and McDowell, J. H. (eds.) *Andean Cosmologies through Time: Persistence and Emergence*. Bloomington, Indiana University Press, 229–266.
Urton, G., 1996. Constructions of the Ritual-Agricultural Calendar in Pacariqtambo, Peru. In Del Chamberlain, V., Carlson, J. B., and Young, M. J. (eds.) *Songs from the Sky*. Bognor Regis, W. Sussex, Ocarina Books, 180–192.
Urton, G., 1999. *Inca Myths: The Legendary Past*. Austin, University of Texas Press.
Urton, G., 2003. *Signs of the Inka Quipu: Binary Coding in the Andean Knotted-String Records*. Austin, University of Texas Press.
Urton, G., 2011. Tying the Archive in Knots, or: Dying to Get into the Archive in Ancient Peru. *Journal of the Society of Archivists, 32*(1), 5–20.
Van de Guchte, M. J. D., 1990. *Carving the World: Inca Monumental Sculpture and Landscape*. Doctoral dissertation, University of Illinois.
Westerman, J. S., 1998. *The Meaning of Machu Picchu*. Chicago, Westridge Publishing.
Westerman, J. S., 2005. Inti, the Condor and the Underworld: The Archaeoastronomical Implications of the Newly Discovered Caves at Machu Picchu, Peru. In Fountain, J. W. and Sinclair, R. M. (eds.), *Current Studies in Achaeoastronomy: Conversations Across Time and Space*. Durham, NC, Carolina Academic Press, 339–351.
Williamson, R. (ed.), 1981. *Archaeoastronomy in the Americas*. Los Altos and College Park, Ballena Press, 305–318.
Woolard, E. and Clemence, G., 1966. *Spherical Astronomy*. New York, Academic Press.
Wright, K., 2006. *Tipon: Water Engineering Masterpiece of the Inca Empire*. Restone, VA, ASCE Press.
Wright, K. and Valencia, A., 2000. *Machu Picchu: A Civil Engineering Marvel*. Restone, VA, ASCE Press.
Wright, R. and Valencia, A., 2001. *The Machu Picchu Guidebook*. Johnson Books, Boulder, CO.
Wright, K. R., Wright, R. W., Valencia Zegarra, A., and McEwan, G. 2011. *Moray: Inca Engineering Mystery*. Reston, VA: ASCE Press.
Zawaski, M., 2007. *Archaeoastronomical Survey of Inca Sites in Peru*. Masters thesis, University of Northern Colorado.
Ziólkowski, M. and Sadowski, R., (1989). Time and Calendars in the Inca Empire. *BAR International Series 479*. Oxford, British Archaeological Reports.
Ziolkowski, M. and Kosciuk, J., 2018. Astronomical observations in the Inca Temple of Coricancha (Cusco)? A critical review of the hypothesis. *TEKA Komisji Architektury, Urbanistyki I Studiów Krajobrazowych, XIV*(1).
Ziólkowski, M., Kościuk, J., and Victoria, F. A. (2013). Astronomical observations at Intimachay (Machu Picchu): A new approach to an old problem. In I. Sprajc and P. Pehani, *Ancient cosmologies and modern prophets* (pp. 391–404). Ljubljana, Slovenia: Slovene Anthropological Society.
Zuidema, R. T., 1964. *The Ceque System of Cusco: The Social Organization of the Capital of the Inca*. Leiden: E.J. Brill.
Zuidema, R. T., 1968. Review of Pablo Joseph De Arriaga and L. Clark Keating: The Extirpation of Idolatry in Peru. *American Anthropologist,* 70, 6, 1224.
Zuidema, R. T., 1977. The Inca Calendar. In Aveni, A. (ed.). *Native American Astronomy*. Austin, University of Texas Press, 219–259.
Zuidema, R. T., 1981a. Anthropology and Archaeoastronomy. In Williamson, R. (ed.) *Archaeoastronomy in the Americas*. Los Altos and College Park: Ballena Press, 29–31.

Zuidema, R. T., 1981b. Inca Observations of the Solar and Lunar Passages Through Zenith and Anti-Zenith at Cuzco. In Williamson, R. (ed.) *Archaeoastronomy in the Americas*. Los Altos and College Park: Ballena Press, 319–342.

Zuidema, R. T., 1981c. Comment on J. H. Rowe: Archaeoastronomy in Mesoamerica and Peru. *Latin American Research Review,* 16, 3, 167–170.

Zuidema, R. T., 1982a. Catachillay: The Role of the Pleiades and of the Southern Cross and α and β Centauri in the Calendar of the Incas. In Aveni, A. & Urton G. (eds.). *Ethnoastronomy and Archaeoastronomy in the American Tropics*. New York, New York Academy of Sciences, 203–229.

Zuidema, R. T., 1982b. The Sidereal Lunar Calendar of the Incas. In Aveni, A. (ed.), *Archaeoastronomy in the New World: American Primitive Astronomy*. Cambridge, Cambridge University Press, 59–107).

Zuidema, R. T., 1983. Hierarchy and Space in Incaic Social Organization. *Ethnohistory,* 30, 2, 49–75.

Zuidema, R. T., 1985. The Lion in the City: Royal Symbols of Transition in Cuzco. In Urton, G. (ed.). *Animal Myths and Metaphors in South America*. Salt Lake City, University of Utah Press, 183–250.

Zuidema, R. T., 1986. The place of the *Chamay Wariqsa* in the rituals of Cuzco. *Amérindia,* 11.

Zuidema, R. T., 1987. Reply to David Dearborn and Katharina Schreiber: Here Comes the Sun: The Cuzco-Machu Picchu Connection. *Archaeoastronomy,* X, 28–29.

Zuidema, R. T., 1989a. A quipu calendar from Ica, Peru, with a comparison to the ceque calendar calendar from Cuzco. In Aveni, A. F. (ed.) *World Archaeoastronomy*. New York: Cambridge University Press, 341–351.

Zuidema, R. T., 1989b. Cuzco, Quipu and Quadrant. In Lavin, I. (ed.) *WORLD ART: Themes of Unity in Diversity. Acts of the XXVIth International Congress of the History of Art*. University Park, PA, Penn State Press.

Zuidema, R. T., 1989c. El Ushnu. *Reyes y Guerreros. Ensayos de Cultura Andina.* Grandes Estudios Andinos. Lima, Fomciencias, 402–454.

Zuidema, R. T., 1990a. *Inca Civilization in Cuzco*. Austin, University of Texas Press.

Zuidema, R. T., 1990b. Ceques and Chapas. An Andean Pattern of Land Partition in the Modern Valley of Cuzco. In Illius, B. and Laubscher, M. (eds.). *Circumpacifica Festschrift für Thomas S. Barthel.* Frankfurt am Main, Peter Lang, 627–643.

Zuidema, R. T., 1991. Review of Michael J. Sallnow: Pilgrims of the Andes: Regional Cults in Cusco. *The Hispanic American Historical Review,* 71, 1, 152–153.

Zuidema, R. T., 1992. Inca Cosmos in Andean Context: From the Perspective of the Capac Raymi Camay Quilla Feast celebrating the December Solstice in Cuzco. In Dover, R. V. H., Seibold, K. E., and McDowell, J. H. (eds.) *Andean Cosmologies through Time: Persistence and Emergence*. Bloomington, Indiana University Press, 17–45.

Zuidema, R. T., 1996. Review of Brian S. Bauer & David S. P. Dearborn: *Astronomy and Empire in the Ancient Andes: The Cultural Origins of Inca Sky Watching. The Americas,* 53, 2, 295–297.

Zuidema, R. T., 2002. Review of Brian S. Bauer: *The sacred landscape of the Inca: the Cusco ceque system. Antiquity*, 76, 292, 592–593.

Zuidema, R. T., 2005. The Astronomical Significance of a Procession, a Pilgrimage and a Race in the Calendar of Cusco. In Fountain, J. W. and Sinclair, R. M. (eds.) *Current Studies in Archaeoastronomy: Conversations Across Time and Space*. Durham, NC: Carolina Academic Press, 353–367.

Zuidema, R. T., 2007. Solar and Lunar Observations in the Inca Calendar. In Ruggles, C. and Urton, G. (eds.) *Skywatching in the Ancient World: New Perspectives in Cultural Astronomy.* Boulder, University Press of Colorado, 269–285.

Zuidema, R. T., 2008a. Pilgrimage and Ritual Movements in Cuzco and the Inca Empire. In Malville, J. M. and Saraswati, B. (eds.) *Pilgrimage: Sacred Landscapes and Self-Organized Complexity*. New Delhi: Indira Ghandi National Centre for the Arts, 269–288.

Zuidema, R. T., 2008b. The Astronomical Significance of Ritual Movements in the Calendar of Cuzco. In Staller, J. E. (ed.), *Pre-Columbian Landscapes of Creation and Origin*. New York: Springer, 249–268.

Zuidema, R. T., 2015. Ceque System of Cuzco: A Yearly Calendar-Almanac in Space and Time. In Ruggles, C. L. N. (ed.) *Handbook of Archaeoastronomy and Ethnoastronomy*. New York: Springer, 851–863.

Zuidema, R. T., and Quispe M. U., 1973. A visit to God – The account and interpretation of a religious experience in the Peruvian community of Choque-Huarcaya. *Bijdragen tot de Taal-, Land- en Volkenkunde,* 124, 1, 22–39.

Index

© Springer Nature Switzerland AG 2020

S. R. Gullberg, *Astronomy of the Inca Empire*, Historical & Cultural Astronomy,
https://doi.org/10.1007/978-3-030-48366-1

The manufacturer's authorised representative in the EU is Springer Nature Customer Service Centre GmbH, Europaplatz 3, 69115 Heidelberg, Germany. If you have any concerns regarding our products, please contact ProductSafety@springernature.com

Printed and bound by CPI Group (UK) Ltd, Croydon, CR0 4YY
30/09/2026
02225559-0017